KB024835

Biotechnology on a Table

Biotechnology on a Table

First Edition **January 23, 2017**

Author **Agricultural Biotechnology Studying Group**

Translator **Transbio**

Publisher **Seon-ki Kim**

Published by **Purengil. co., Ltd**

Registration Number **16-1292**

#1008 Daerung Post Tower 7. 48, Digital-ro 33gil, Guro-gu, Seoul, Republic of Korea

E-mail **purungilbook@naver.com**

Homepage **http://www.purungil.co.kr**

Phone **+82-2-523-2907**

Fax **+82-2-523-2951**

Copyright © 2009 Agricultural Biotechnology Studying Group

English translation copyright © 2017 by Purengil Co., Ltd.

Printed in Korea

ISBN 978-89-6291-379-8 03470

국립중앙도서관 출판예정도서목록(CIP)

Biotechnology on a table / author : Agricultural Biotechnology Studying Group ; translator: Transbio. -- Seoul : Purengil, 2017
p. ; cm

원표제: 식탁 위의 생명공학
Translated from Korean
ISBN 978-89-6291-379-8 03470 : ₩15000

521.7-KDC6
630.27-DDC23 CIP2017001011

Correct understanding of biotech crops

Biotechnology on a Table

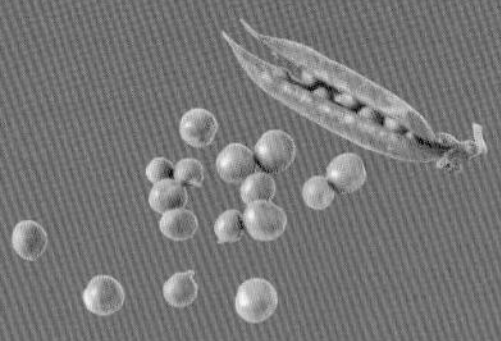

Agricultural Biotechnology Studying Group

•

Translation Transbio

PURENGIL

Contents

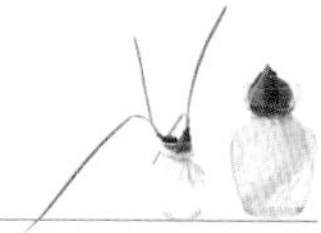

Preface

It is predicted that, in the 21st century, arable land will decrease and agricultural environment will become impoverished due to the accelerated industrialization accompanied with a population explosion. Even now, exhaustion of fossil energy resources wrought the development of alternative energy to be urgently required, and the earth calls for environment conservation stronger than in any other times. In short, foods, energy resources, and environmental problems are all pressing matters that we should solve in the near future. Hereupon, the utilizationof genetically modified crops is suggested in the field of science, as a means of preserving the environment and augmenting supplies of food andalternative energy resources. Developed countries are making massive investment in plant genomics research and, governments nurture plant biotechnology industry strategically.

The methodology to make genetically modified crops (GMOs), a technology that transfer plant's desirable traits by the expression of transplanted foreign genes, was established with the commercialization of FLAVR SAVR® (a delayedripening tomato) in 1994. Since then, numerous crops such as herbicide-tolerant soybeans, Canola, cotton, insect-resistant corns have been developed and commercialized. By 2008, 132 events of 22 crops had been commercialized. As such, GM crops are evolved into 'designer plant' technology, being established in our everyday lives. in addition to the current pest/disease-resistant GMOs, new varieties of high value-added crops for functional foods and neutraceuticals with health benefits will be consistently developed and commercialized in the next generation.

However, doubts concerning environmental and food safety of GMO were raised, and production and consumption of GMO are going through a trial. To meet such damands, government of individual country is providing regulatory protocols to make sure safety of GMOs, and scientists are putting their best efforts to development safer products.

By providing scientific knowledge and principles on GMO technology, this book would like to contribute to understand the process of the development, safety and possible risk of its output. Authors of experts working in various fields of development and regulation, tried to provide up-to-dated information about genetically modified agricultural products on the basis of science.

Scientific technologies are double-edged: they contribute to convenient lives of human beings, but at the same time they may carry a chance of risks. It is just like as we cannot give up the convenience of the automobile provide to us just because it pollutes the air and damages people by traffic accident. In fact, technology itself is neither good nor bad. It depends on how we use the technology.

Even after the automobile had been gener ally used as a public transportation, its speed was limited upto 4 mph (6.4 km/hr) by 1896. speed limit in city area was 2 mph (3.2 km/hr), which is even worse. Aside from speed regulation, it was required to put notable signals (red flags during day, and red lamp during night) in front of the car in order to prevent any accidents. Some would say that excessive regulations had retarded the development of automobiles. However, authors think that new technologies have always started as such, and GMOs are no exception.

Positive and negative aspects of a scientific technology should be dispassionately examined based on the situation of the community, and consumers should make the ultimate choice. Making choice and purchasing are the rights of consumers. Governments should evaluate potential risks and benefits, and consider appropriate regulation costs for risk assessment of biotechnological crops. Consumers should trust the legal measurement of their government.

Consumers, also, should actively take an interest on the new technology and acquire knowledge on GMO in order to make a right decision. Only when people understand the basic principles of a scientific technology will they be able to resolve moral, social, and economic issues, and relieve anxiety coming from vague fear and indifference. Along with the development of biotechnology, the general public will have to make biotechnological decisions every moment: not only on their diet, but also on health care including medical treatments for disease, and even the genotypes of their offsprings.

We give thanks to everyone who spared precious time to read the copy of this book, and especially appreciate the members of 'Agricultural Biotechnology Study Group'. Hope that this little book can contribute to the development of biotechnology.

Biotechnology on a Table

Chapter One

Biotechnological Crops

For a person living in the 21st century, words such as biotechnology, genetic engineering, DNA recombination, genetic modification or GMO (genetically modified organism) are probably very familiar. Watching TV or reading newspapers every morning, one could see how busy they are reporting rapid development of biotechnology. But when it comes to the story of human cloning, the discussion halts with anxiety. It is miserable when discussions about GMOs, which are especially prevalent, mostly end with negative conclusions. GMOs are so common around us that among the imported agricultural products in 2007, 79% were herbicide-tolerant soybeans and 51% were insect-resistant corns. Nevertheless, controversies about them are endless.

Development of biotechnology resulted in extension of human lifespan, and consequent increase in population predicts severe food shortage in near future. A solution is required to reduce environmental pollution and contamination of agricultural products. Among various plans to resolve the problem, GMOs and foods made from them are being emerged as likely alternatives. New agricultural biotechnologies that can help produce clean products are required. However, many people do not thoroughly understand the concept of GMO and just take a negative stance groundlessly. 21st century is the era of biotechnology. From now on, various forms of GMOs will be continuously developed and placed in the market. Therefore, we, the consumers, need scientific understanding of what GMOs are, how they are made, and whether they are actually safe.

1-1. Conventional Crossbreeding and the Green Revolution

Traits or characteristics of all organisms on earth are determined by genetic information in their cells. Each unit of this information is called a gene, and it exists in a cell in the form of DNA. That is why an entity's characteristics change along with the change of its genetic information. This also accounts for the reason why even family members, you, your parents, and your siblings differ, in many aspects, from one another. Just as siblings who are born from the very same parents are not the same, all existences differ from one another depending on the degree of genetic information recombination. In the process of reproduction, DNA is so diversely recombined that the offspring produced as a result greatly vary in their traits.

Meanwhile, diversity of flowers is more often obtained by mutation that occurs in natural condition than by the genetic recombination that occurs during reproduction. The range of fertilization carried out by the combination of genetic information is very confined. Such a phenomenon results from the long evolution process which has confined the possible range of breeding into only between the same

Figure 1-1. Tall wheat and dwarf wheat
The wheat in the right side of the picture is the tall wheat produced before the Green Revolution and the left one is the dwarf wheat that had gone through the Green Revolution.

Figure 1-2. Power of crossbreeding. As a result of repetitive crossing for 5,000~10,000 years, humans finally obtained corns of today's form (right) from wild teosinte (left).

species; it is inevitable that the diversity of genetic information is restricted. This phenomenon is shown by the inability of two different species to fertilize with each other, as in between a horse and a cow. For this reason, humans have diversified genetic information of many species by a series of process called crossbreeding: the process of storing a fraction of seeds of good crops and planting them the next year, or breeding good seeds to produce better crops.

The magnificent effect of cross-breeding is very well-known via the Green Revolution in the 1960s. Breeding short crops from originally tall crops like wheat, rice, corns etc, energy that had previously been expended to grow the long leaves and stems was reduced. In addition, due to their shorter height, the newly bred crops were less vulnerable to storms and winds, a trait that made it possible to attach large spikes to them (Figure1-1). The dwarf gene, doubling the crop production, made the greatest contribution to the Green Revolution, and it had freed then undeveloped countries like Mexico and Pakistan from poverty and hunger. Doctor N. Borlaug, who discovered and further utilized the gene, received the Nobel Peace Prize in the year of 1970.

According to a recent research, this dwarf gene is the one that determines biosynthesis or sensitivity of gibberellins, the plant hormone. The research also revealed that some typical examples of such type of genes are *GAI* in thalecress, *d8* in corns, and *Rht-B1/D1* in wheat. Corns, in fact, had been originally mere weeds, but have become today's corns by 5000~10000 years of breeding (Figure1-2).

In retrospect, rice, not very different from grasses, was also not suitable to eat, especially because of its shattering property. However, ever since the discovery of a mutant with non-shattering property (*sh-2*), it's importance as an agricultural product has been highly recognized, and as a result of long process of breeding, today's rice has been obtained. Even though such a significant discovery and development had taken place, there is still another problem facing. It is predicted that we will experience serious shortage of food within next 30 years. Therefore, it is highly urgent to get high-yield effect equivalent to that of crossbreeding of those days.

1-2. Emergence of Biotechnology

As a result of the long process of breeding, the diversity of female flowers and male flowers, which provide new genetic information, has been exhausted. By the late 1980s, the rate of increase of agricultural productivity had notably decreased. After the 1990s, world's rate of increase in crop production began to fall behind of population growth rate. Hence, a new approach to increase agricultural productivity was required, and the new biotechnology and genetic engineering techniques were applied to conventional method of crossbreeding.

Whereas conventional breeding technique utilizes broad genetic

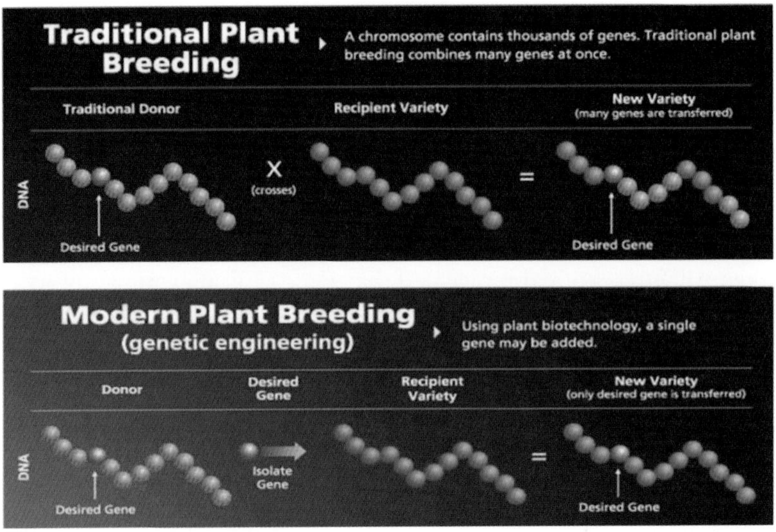

Figure 1-3. Comparison between crossbreeding and molecular breeding by biotechnological method

In conventional breeding recombination of broad range of genes is inevitable (top), but biotechnological method transplants highly limited number of genes so that characteristics of a crop are made exactly as intended (bottom). Note the relative amount of genes being transformed in each case.

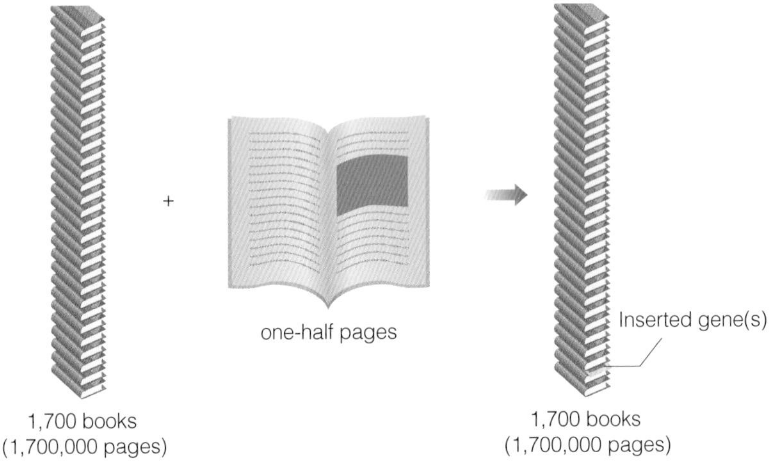

Figure 1-4. Degree of gene transformation in biotechnological crops

If genetic information of a corn that determines its characteristics were to be written down, it would be equivalent to the amount of 1,700 copies of 1,000 page Seoul city phonebooks. But if a new single gene is transplanted by biotechnology, the amount of information transplanted is equivalent to only a half page of a phonebook.

recombination, hybridizing genetically similar organisms and selecting a few superior ones, biotechnological method, molecular breeding, transplants small amount of certain genes, instead of an organism itself, to alter relatively restricted amount of genetic information (Figure 1-3).

Biotechnology allows isolation of genes of interest from almost any organisms and transplant those genes in another organism. The amount of genetic information in a corn cell is about 2.5×109bp, which corresponds to the amount of information in 1,700 phone books, each with 1,000 pages, of Seoul City, as described in Figure 1-4. Still, this is just a small change compared to the broad genetic recombination occurring in conventional breeding process (Figure 1-4).

1-3. Nomenclature of Biotech Crops

Organisms with altered traits from those of the original ones due to gene transplantation are called transformants by scientists. If such organisms are crops such as corns, wheat, and rice, those are called transformed crops. Because genetic information of the organism is partly transformed as a result of gene transplantation, they are also called GMO (Genetically Modified Organism); crops of such kind are called GM (Genetically Modified) crops. In reference to the technique used in generating such crops, they are also called recombinant organisms or recombinant crops.

At times, they are called by the misnomer, genetically 'manipulated' organisms. This term is derived from Japanese term of GMO. Whereas the word 'manipulate' has a neutral meaning in Japanese, however, the same word in Korean is mostly used in negative connotation, and therefore it is not appropriate to be used as a standard term.

As such, calling on the same concept by varying names is likely to wrought confusion and thus not desirable. Even different government departments call them by different names. For example, Korean Ministration of Food, Agriculture, Forestry, and Fisheries call them genetically modified agricultural products, but Korean Foods and Drug Administration call them genetically recombinant foods.

Meanwhile, what becomes a problem in an international trade of these products is when they are alive. For this reason, those kinds of live organisms are specified by the name LMO (Living Modified Organism) or LM (Living Modified) crops. In short, GMO is a broader concept than LMO, and includes it.

Since such terms are not yet standardized, this book calls all such kinds of crops as biotechnological organisms or biotechnological crops. However, since terms used within laws and regulations are different depending on government departments in charge of, we used terms as they are used within laws and regulations in this book.

1-4. Benefits and Safety of Biotech Crops

Previously, crop yields have been increased by using pesticides to decrease the damages from weeds, pests, and diseases, and fertilizers to improve the productivity of soil. Now, biotechnological crops have greatly resolved the stubborn problems of weeds, pests and diseases, from which most of the farmers had suffered. Because the world population is exploding, the development of biotechnological crops is becoming another agricultural revolution following after the Green Revolution (Figure 1-5).

Accelerated industrialization accompanied with population explosion resulted in the reduction of arable land and aggravation of agricultural environment. Furthermore, exhaustion of fossil energy resource wrought the development of alternative energy to be ur-

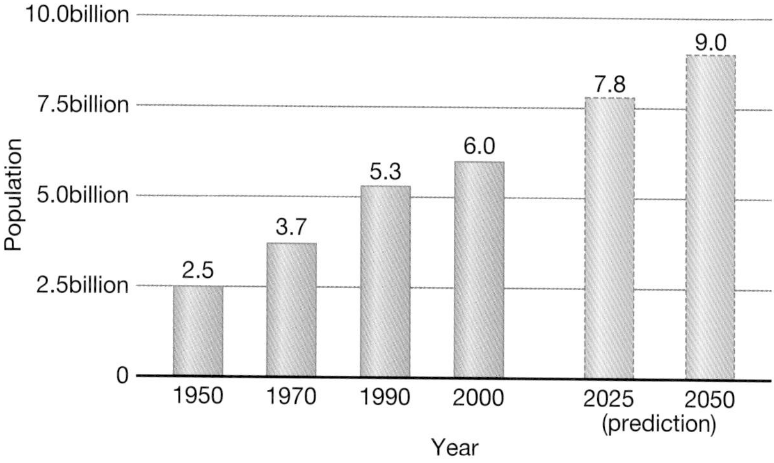

Figure 1-5. World population growth
Current amount of crop yields cannot feed all the world population in the future. In 2030, when world population is predicted to be 8 billion, we will probably be confronted with serious food shortage.

gently required, and the earth calls for environmental conservation stronger than ever. Also, to deal with the worthening agricultural environment, development of eco-friendly alternative technology is definitely required. Therefore, developed countries consider development and utilization of biotechnological crops as the means to increase food and alternative energy resource, and spare no expenditure in plant research, making a massive investment in the field. They are particularly fostering plant biotechnology as the government-supported industry stategically. Along with such efforts from government, genome projects carried on since the late 20th century has made possible the researches to investigate the functions of a genome (an entire structure of genes that decides overall characteristics of a plant) and those of each individual gene included in the structure.

Gene recombination has long been a phenomenon that naturally occurs as a process of life and evolution. Therefore, foods made from biotechnological crops are supposed to have no practical difference from conventional foods chemically, structurally, and biologically. Nations that have developed and commercialized numerous biotechnological crops, such as USA and Canada, are especially confident with their view of substantial equivalence,* which asserts that biotechnologically produced foods are equally safe as the foods produced by conventional methods, according to results from thorough scientific evaluation.

Furthermore, proteins made by the genes of a given GMO should be analyzed by certain processes such as heating, digestion by intestinal and gastric juice, allergenicity test, and should be compared with traditional crops to test if there are any alterations in constituents before it is tested for its safety and is commercialized. Also, it has to meet various requirements from safety standards of each government department in order to be approved and distributed in the market. Because of such complexity, developer of the product has to collect scientific evidence that it is not harmful for both human body and environment through years of safety evaluation procedure and then should submit the results to department in charge of. Considering the fact that food products that we usually consume without any doubt are normally distributed in market without any safety evaluation procedure, biotechnological crops are safer than any other foods in the market.

---

* substantial equivalence: a scientific approach to the evaluation of safety of genetically modified organic in which characteristics of foods that have been empirically determined to be safe are set as the safety level, the evaluation standard, and compared to components (nutrients, nutrition suppressing factors, toxin) and ingestion tendency or predicted amount of intake of a genetically modified organic and analyzing the difference. According to this view, if there exist no risks to the difference, the evaluated food can be regarded as safe.

Nowadays, agricultural biotechnology is being considered as a powerful alternative that could not only increase productivity, preserve environment, and contribute to the improvement of safety and quality of food products, but also increase competitiveness of agriculture. Development of biotechnological crops is currently led by developed countries like USA and Canada, but since there are not much gap between technologies of such nations and those of our country (The authors of this book is South Korean), it is recognized that plant biotechnology in Korea is also a promising field of industry. Considering current situation of our country, where we depend on import of most of the crops and self-sufficiency is very low, we should never overlook the merits of biotechnological crops.

While one side of the ongoing debate on commercialization of biotechnological crops is constantly arguing for the potential risks in them, the research results that claim the risks of those products have never confirmed reproducibility. According to scientific perception, it is the general consensus from scientists that biotechnological crops are safe. Humanity adapts to and uses contemporary technologies like electricity, automobiles, and airplanes despite their potential risks because advantages of using these far more outweigh the risks. Therefore, both government and the people, at the same time regarding safety as the most important factor, should learn to have balanced view and properly regulate biotechnological crops so that healthy research and development is not hindered.

By 2008, 144 events of 24 different biotech crops including soybeans, corns, squash, rapeseed, cotton have been commercialized. In addition to herbicide-tolerant and insect-resistant biotech crops, designer crops with improved functionality or increased nutritional or medicinal components are being developed and commercialized recently The most typical example would be Golden Rice® 1 and

2, developed in 2000 for the first time by increasing the amount of vitamin A precursor in normal rice, and Vistive®, a new cooking oil with ω-3 fatty acid and lower trans-fat developed by changing the composition of unsaturated fatty acid. It is perceived that in the near future, various crops, such as tomatoes high in hepatitis vaccine, will be on our table. Industrial competitiveness will be determined by the uniqueness of useful genes to be utilize in biotech crops. Developed countries accelerate the discovery of useful genes with the support at national level, whereas in our country, it has been stagnant with only a little support from government and private sectors. Therefore, we are currently at a difficult situation in which there are hardly any useful genes and transformation technology, the key factors to strength competitiveness. However, authors expect that such difficulties could be overcome by the large-scale research support recently led by Korean government.

Ever since some scientists began to make an issue of potential environmental risk of biotechnological crops, major biotechnological crops-producing countries have put their endeavors into improving risk evaluation technology and into organizing state regulatory system in order to prove safety of biotechnological crops. Several countries like EU nations and Japan are enforcing labeling system, setting up a certain limit on the contamination of biotechnological crops. In the case of South Korea, the range of labeling system has been expanded from only a certain portion of biotechnological crops to all of them. However, this system is not the indication of risks of biotechnological crops. You will be able to understand from this book that all such kinds of system is just to satisfy consumers' rights to be informed.

Chapter Two

History and Status Quo of Biotech Crops

2-1. History of Biotech Crops

After obtaining the result that agrobacteria, a kind of pathogenic bacteria, induces tumor in plants, scientists became aware that part of the bacterial genes are transformed to plant cells and then expressed. Chilton, Fraley, Schell and etc. succeeded in proving that if the bacterial genes are substituted by useful genes, the plant's genetic information is altered and thus can genetically transform it.

History of development of biotech crops started in 1983, when kanamycin (an antibiotic)-resistant tobaccos and petunia were first nurtured. Following after Belgium, in 1986 USA and France, the packing experiment for herbicide-resistant cigarettes was officially carried out. In 1987, cotton was transformed and in 1988, transformation of beans and rice took place. In 1992, wheat was transformed. Meanwhile, in 1990, packing experiment for blight-resistant cotton was successfully carried out.

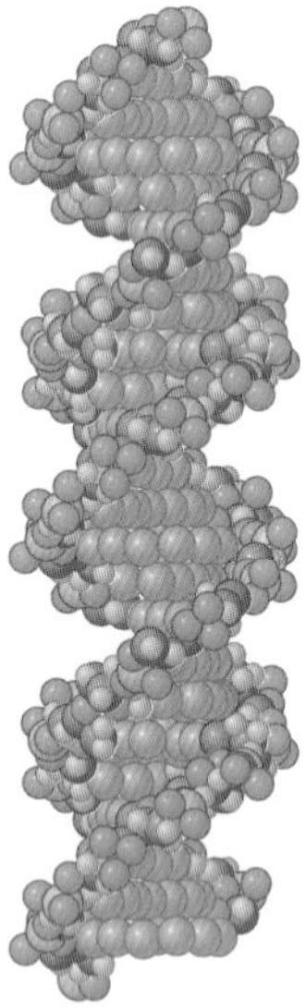

1953	Investigation of structure of DNA
1973	Development of recombinant DNA technology
1977	Development of DNA base sequencing method
1983	Development of herbicide-resistant tobacco
1986	Field test for herbicide-resistant tobacco
1994	Commercialization of delayed-ripening tomato, FLAVR SAVR™
1995	Commercialization of Herbicide-tolerant soybean, insect-resistant corn
2000	Development of Golden Rice™
2008	Commercialization of 144 varieties of 24 biotech crops

Figure 2-1. History of development of biotech crops

In 1994, Flavr Savr® (delayed-ripening tomato) was first commercialized and introduced to the market. Starting from this event, in 1995, Monsanto released beans that do not react to Roundup®(the company's widely-known herbicide) with microorganism enzyme gene transplanted in them. In addition, starting from 1996, large-scale commercial cultivation began to prevail, along with actual commercialization of YieldGard®, insect-resistant corns and Bollgard®, insect-resistant cotton (Figure 2-1).

In the year of 2000, Golden Rice with strengthened vitamin A was developed and was distributed for free to underdeveloped nations, and analysis result of dielectric of rice was freely shared by scientists all over the world, an event of monumental significance. By 2008, 132 events of 22 biotech crops were permitted for commercialization.

Although most of the items of biotech crops are still only herbicide-tolerant or insect-resistant, various kinds of crops, such as the crops with improved quality, are under examination.

Table 2-1. Status quo of currently ratified items of biotech crops (Agbios, 2008)

Crop	Genetically modified trait (number of items ratified)
Potato	Insect-resistance(2), Insect/virus-resistant(2)
Tobacco	Herbicide-tolerance(1), Reduced nicotine(1)
Lentil	Herbicide-tolerance(1)
Melon	Controlled-harvest maturity(1)
Cotton	Insect-resistance(6), herbicide-tolerance(5), Insect-resistance/ herbicide-tolerance(7)
Wheat	Herbicide-tolerance(7)
Rice	Herbicide-tolerance(5)
Sugar beet	Herbicide-tolerance(3)
Squash	Virus-resistance(2)

Argentine canola	Herbicide-tolerance(7), male-sterility/herbicide-resistance(5), improved fatty acid(3)
Flax	Herbicide-tolerance(1)
Alfalfa	Herbicide-tolerance(1)
Corn	Insect-resistance(7), herbicide-tolerance(9), insect-resistance/herbicide-tolerance(23), male-sterility/herbicide-tolerance(3), enhanced lysine(1), enhanced lysine/insect-resistance(1), enhanced amylase(1)
Chicory	Male-sterility/herbicide-tolerance(1)
Carnation	Life extension/herbicide-tolerance(1), changed flower color/herbicide-tolerance(2)
Soybean	Herbicide-tolerance(7), enhanced oleic acid(2)
Creeping bentgrass	Herbicide-tolerance(1)
Tomato	Controlled harvest maturity(5), insect-resistance(1)
Papaya	Virus-resistance(1)
Canola	Canola herbicide-tolerance(2)
Plum	Virus-resistance(1)
Sunflower	Herbicide-tolerance(1)
Total: 22 kinds of crops	132 items

2-2. The Status Quo of GMO Crop Yield

GMO crops have been cultivated since 1996, starting in the USA. Now, they are cultivated worldwide in 25 countries including the USA, Argentina, Canada, China, and consumed by 55 countries.

The cultivation area of these crops has been growing continuously since 2000, by 14% per year. In 2007, GM crops were cultivated from 125 million ha of arable land, which is a 9% rise compared to farming area of the previous year. This area accounts for about 8% of the total cultivation area and 60 times the area of agricultural arable land in South and North Korea (Figure 2-3).

In 2008, 13.3 million farmers from the 25 countries cultivated GM crops. If you look at cultivated area of each country, America culti-

Figure 2-2. The status quo of GM crop consuming country
Countries in green are the 25 countries that cultivate GM crops. Countries in yellow are the 30 countries that import GM crops.(ISAAA, 2008)

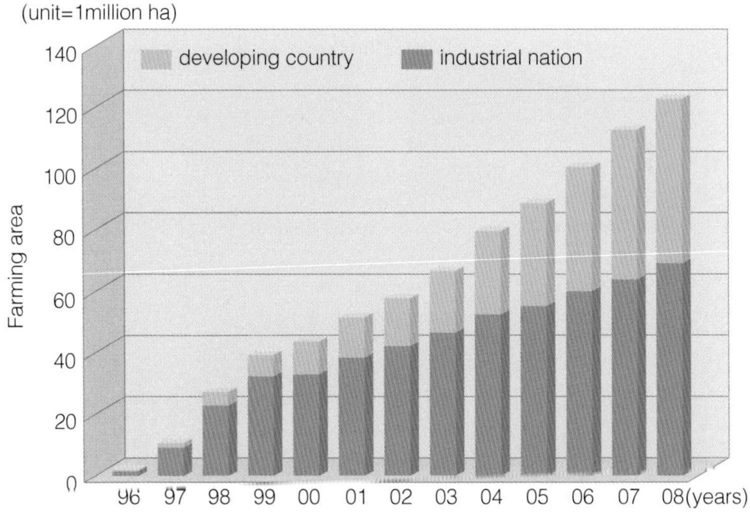

Figure 2-3. Area of GM crop fields over the years (ISAAA, 2008)

vated GM crops in 62.5 million ha of land, which is about 50% of the worldwide cultivation land (Figure 2-4). Argentina cultivated 21 million ha (17%), Brazil, 15 million ha (13%), Canada, 7.6 million ha (6%), China, 3.8 million ha (3%), of GMO. About 43% of the GMO cultivation area, amounting around 54.7 million ha, is currently dominated by third world countries, and has been continuously rising.

By crop, soybeans are cultivated in 65.8 million ha of land worldwide, and take up 53% of the total cultivation area (figure 2-5). GM corns are cultivated in 37.3 million ha (30%) of land, which exceeds 35.2 million ha of 2007. Cotton takes up 15.5 million ha (12%) of

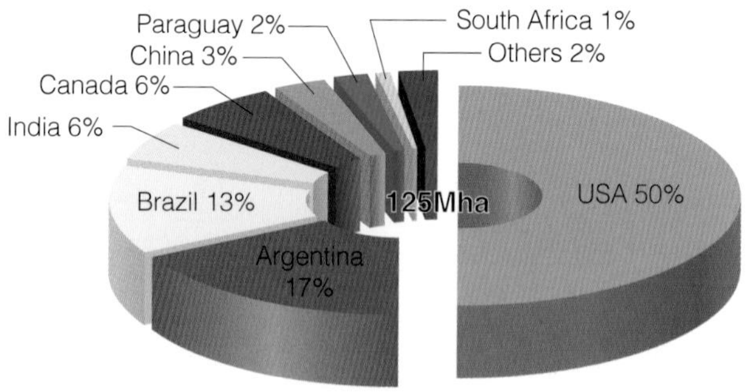

Figure 2-4. Relative areas of major producing countries (ISAAA, 2008)

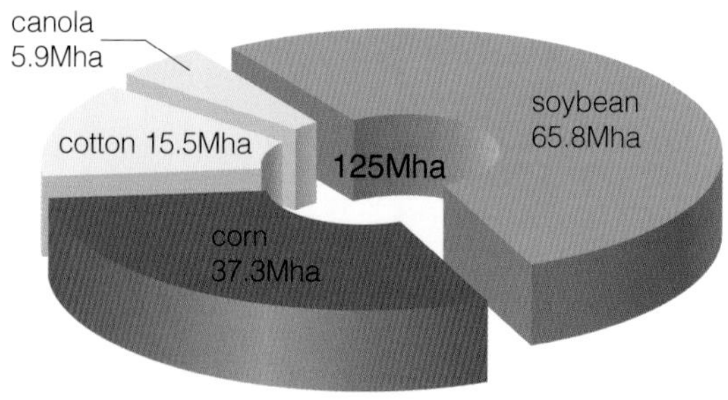

Figure 2-5. Cultivation area per GM crop (ISAAA, 2008)

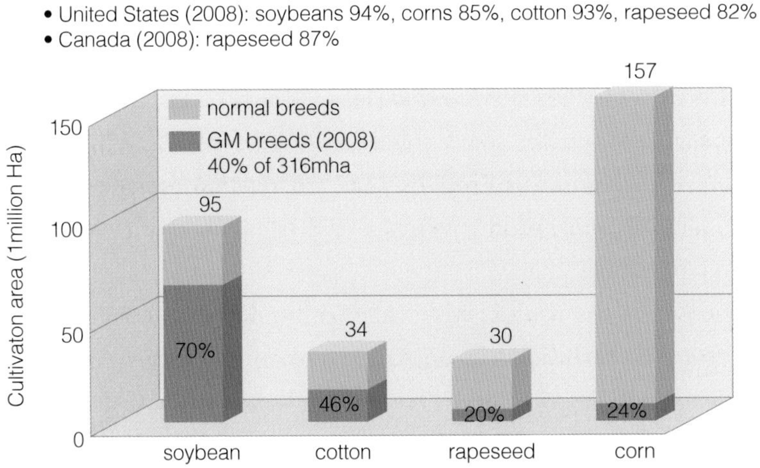

Figure 2-6. Cultivation area and portion of major biotech crops

land, which is an increase from the previous year's 15 million ha, and rapeseeds have been cultivated in 5.9 million ha (5%) of land.

In the case of GM soybeans, they are being cultivated 65.8 million ha of land and take up 70% of the total soybean cultivation (Figure 2-6). GM corn is cultivated in 37.3 million ha of land, taking up 24% of total corn cultivation land, GM cotton in 15.5 million ha, which is 46% of total cotton cultivation area. GM canola in 5.9 million ha, taking up 20% of total cultivation. Out of these four crops, 40% on average is being cultivated as GM crops.

In the case of the US in 2008, 94% of cultivated soybeans are GM crops, corns 85%, cottons 93%, canola, 82% (Figure 2-6). Especially the US, Argentina, and Brazil, the great agricultural countries, have been quickly switching to GM crops and have well passed the world-wide average. We can easily imagine the difficulty of importing non-GMO crops from these countries, and the extra cost that may come along with it. In the US, its grocery stores' products are 80%

GM crops or food based on GM crops. GM crops are can easily be seen in the states. Even so, the GM crops are not specially labeled or separately sold. They are just sold the same way other crops are sold. Canada is the number one canola exporting country, and in 2006 to 2007 the cultivation area rose by 15%. Current GM canola takes up 90% of total canola production, and 62% of canola oil is being made from GM canola.

Looking at the transformed traits within soybean, corns, canola and cottons, herbicide-tolerant crops took up 63%, insect-resistant Bt crops took up 18%, and 19% of the crops were both herbicide-tolerant and insect-resistant (Figure 2-7).

The genetic modification technology not only contributed academically in molecular genetic studies, but also created a new business market of practical GM crops. Since 1983 with the creation of the first GM tobacco, the US, in 1994, first cultivated FLAVR SAVR®, delayed ripening tomatoes, and in 1996 cultivation of GM crops in large quantities began. Now soybean, corn, cotton and canola that are herbicide-tolerant and insect-resistant, along with 24 crops and

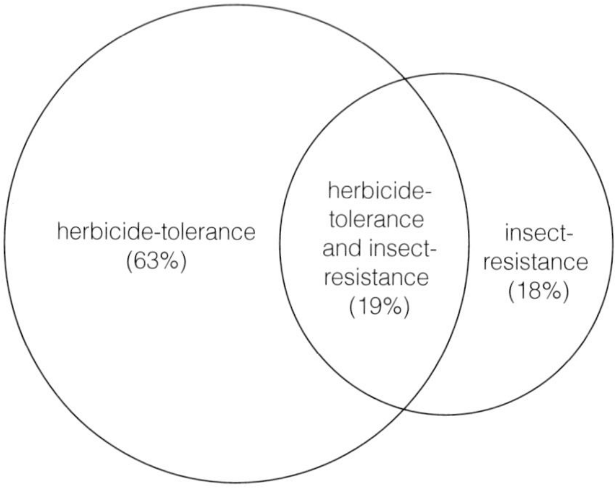

Figure 2-7. Distribution of transformed traits

144 items in total are either commercialized, or are in the process of being commercialized. However it is rather limited to dicotyledons such as soybeans, canola, tomato, and potato. Development of major cereal crops that are monocotyledons, such as rice, wheat and corn, are very limited.

Biotechnology on a Table

Chapter Three

Processes of Biotech Crops Development

Genetically modified crops are generated through crossing over species by transformation of crops with genes of desirable traits extracted not only from plants but also from animals and microorganisms. Let's look into the processes that these genetically modified crops go through.

3-1. Extracting Genes of Desirable Traits

An organism's characteristics are determined by its genes. In this book, we define agriculturally useful genes as desirable gene. There are billions of stars in night sky but not one is without a name. However, in the natural world, out of all the strings of genetic information out there, genes that have its functions experimentally proven are less than 5%. Many biologists are trying to discover a gene with a specific role. Once we discover the function of a gene, utilization of the gene to change the organisms' characteristic is no longer a novel technology, and this book will explore real examples. The problem is that we do not know which gene determines certain characteristic. Due to the development of biology and related fields, uncovering specific gene will be accelerated. However, still finding a big number of genes and discovering what role it plays will be a task for the next generation of biologists.

3-2. DNA Recombination

Desirable genes that aim transformation are first recombined in a test tube so that they can function properly inside a plant. For a desirable trait to be shown properly in a plant cell, the promoter and the terminator, which control gene expression, are recombined with the coding sequence of gene of desirable trait (Figure 3-1).

By using the appropriate promoter, gene's expression rate, timing and expression site can be controlled. The most commonly used

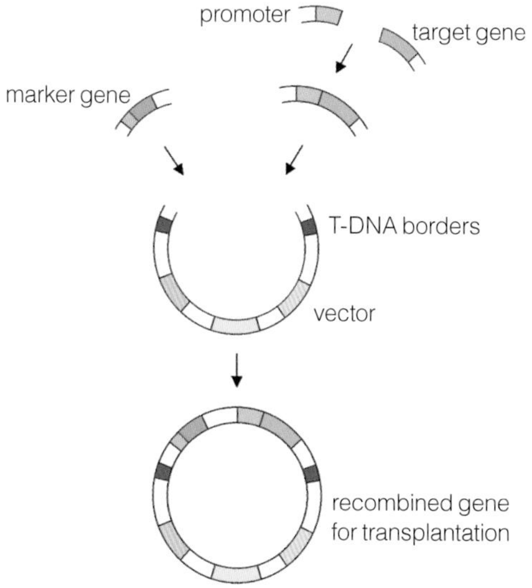

Figure 3-1. Agrobacterium-mediated genetic recombination for transplantation of target gene

generic promoter in biology is the CaMV 35S promoter, a powerful promoter from the Cauliflower Mosaic Virus. By recombining this promoter and useful gene, extensive expression all over the organism in growth periods can be possible, but is expressed partially as well in some cases. By using promoters that only work on leaves or roots, for example, specific parts of plants are being damaged can be protected effectively. Or else it can go the other way around. If the part of the plant that is being consumed is the seeds or the fruits, the effect of the transplanted gene can be minimized in those organ, so the amount of protein consumed can be lowered. However, most genes need to be expressed differently according to stages in development and differentiation, so the use of effective inducible promoters is desirable.

On the other hand, by reversing the orientation of gene but using the same DNA, genes can be silenced using RNA inhibition. Thus, by activating the target gene's anti-sense mRNA, it will either suppress the expression of sense RNA by repressing protein translation or just destroy the sense mRNA. By using this technology that effectively repress genes that were originally in organisms, products such as rice that doesn't induce allergies, tomatoes that do not go stale easily, coffee without caffeine, and tobacco without nicotine were invented.

In the process of DNA recombination, selection marker genes are inserted, in order to select cells that were genetically transformed. Selection marker genes are inserted to degrade herbicides and antibiotics, so that the gene would improve resistance. When selection marker genes are recombined with useful gene, in case the organism is grown in a setting with those antibiotics or herbicides, only the ones with resistance would survive, making the selective process much easier, thus making it easier to reproduce.

The selection marker genes that are used for dicotyledons are usually *neomycin phosphotransferase* to give resistance against antibiotics kanamycin. Selection markers that are mostly used for monocotyledons are *hygromycin phosphotransferase gene* (*hyg*), which confers resistance against antibiotics hygromycin, or *phosphinothricin acetyltransferase genes* (*pat or bar*), which gives resistance against herbicides. When a herbicide-resistant gene is used, herbicides could be used to make the selective process easier in the field to deal with a large number of plants. Also, the resistance against herbicides itself gives an advantage agriculturally. The *bromoxynil nitrilase gene*, *acetolactate synthase gene* are also used as herbicide-resistance genes.

Sometimes we use reporter genes that are handy indicators that tell whether the transformed gene was activated or not. Examples of these would be the enzyme gene *β-glucuronidase*, the green fluorescent protein gene, and firefly's light-emitting protein gene (Figure 3-2).

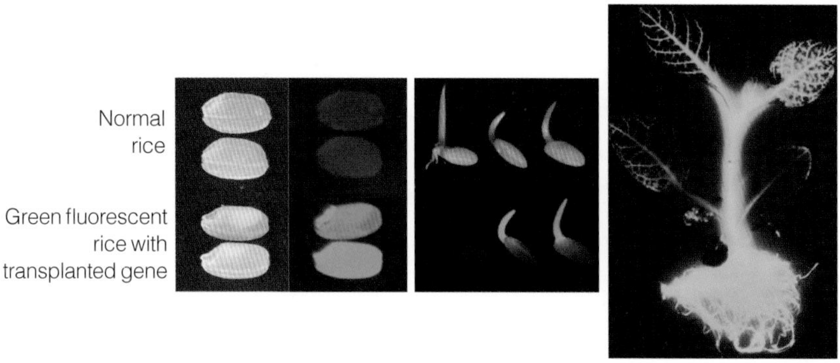

Figure 3-2. Rice that had been transplanted Green Fluorescent Protein (GFP) gene

Agrobacterium-mediated transformation is the most popular method to change the characteristics of plants, as shown in Figure 3-1. A 25 bp of the T-DNA border sequences are recombined with useful gene and selection marker. In the Agrobacterium method, the genes surrounded by the border sequences would be inserted into the chromosome of the plant. Organisms with modified traits are the result of these various attempts to transform useful gene into plant cells.

3-3. Methods to Transform Recombinant Genes

Common ways used to transform recombinat gene into the crops include the use of *Agrobacterium tumefaciens*, a germ found in nature, and the microprojectile bombardment, in which gold particles are coated with the recombined DNA and bombarded onto plant cells.

A case of foreign genes being transformed into the nucleus of a plant cell can be found from the germ *Agrobacterium tumefaciens* in nature. When there is less nutrients in the soil, it has the tendency to move into the plant and be parasitic on plants. When it moves in, it transforms a part of its gene to the plant organism so it would produce the necessary nutrients it needs. When unnatural bumps-called crown gall* are made in the plants, it is because the moved *Agrobacterium* gene orders the plant to produce opine, an amino acid derivative, sustenance for the Agrobacterium (Figure 3-3).

Figure 3-3. Crown gall
Carcinoma developed in the agrobacterium-infected part of a plant stem.

Like this, Agrobacterium has the amazing talent of transforming its own genes into that of a plant. The gene piece in an *Agrobacterium*'s Ti plasmid,** T-DNA, defined by the 25 bp long, starting '*BR*' part and the ending '*BL*' part, gets inserted into the chromosome of the plants. Biologists who understood this process cut out opine gene between the *BR* and *BL* and inserted useful gene instead inside the *Agrobacterium*, and developed a way to transfer any gene we want into the plant (Figure 3-4).

With the microprojectile bombardment, the recombined DNA is coated on 1~2 μm of metal particles and shot into the plant cells (Figure 3-5). The metal used here is either tungsten or gold particles.

The foreign gene imported into nucleus with the metal gets in-

* crown gall: a carcinoma developed in the diseased part of an organism. Light gray or lemon yellow-colored and appears as a round-shaped bump at first, but its tissues are soft. Could be developed any time during entire period of growth and development, and is especially likely to appear in hot and humid conditions. Mostly occur at roots or branches.

** Ti plasmid: Huge plasmid of *Agrobacterium tumefaciens* needed to create tumor in infected plant. Modified Ti plasmid is used as a carrier to introduce foreign DNA into the plant's cells.

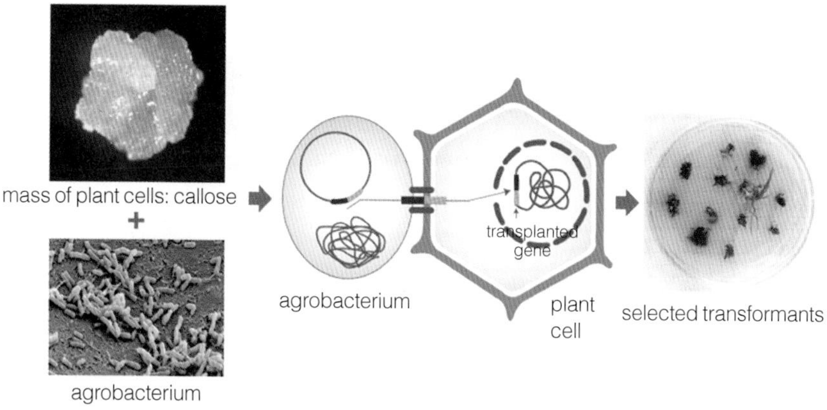

Figure 3-4. The principles by which a gene is transplanted by an agrobacterium

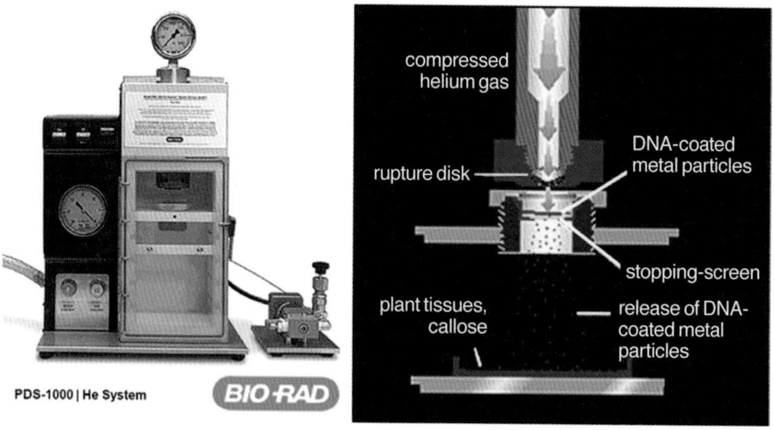

Figure 3-5. Microprojectile bombardment and the internal structure of the instrument used

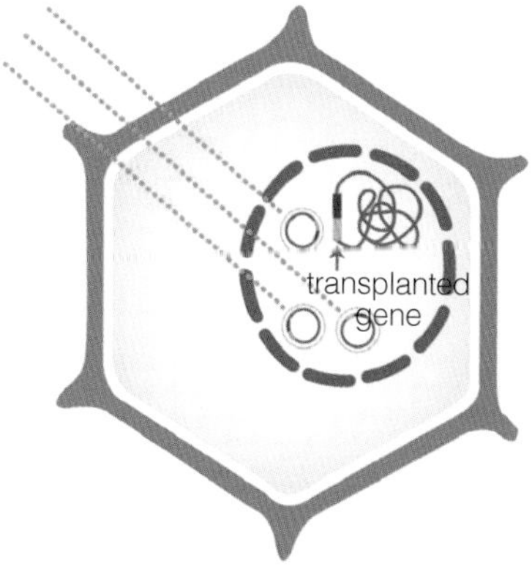

Figure 3-6. Transplanting foreign genes into a plant cell using micropojectile bombardment

serted into the plant genes and is expressed normally and is passed onto the next generation via the Mendel's law. The microprojectile bombardment has an upper hand in which multiple genes can be transformed together. Useful genes and selection marker genes can each be inserted into different chromosomes. By this method, in the passaged culture, the selection marker can be separated easily.

Other than this method, many methods to insert foreign genes into crops are being developed according to the traits of each plant.

3-4. Regeneration and the Event Selection of Biotech Crops

After recombinant genes are transformed into plant cells using selection marker genes such as antibiotics or herbicide-resistant genes, transformed cells can be grown in selection medium with selective agents such as *kanamycin* so that only cells transformed with selection marker gene would survive. Through the regeneration process on the proper nutrients, a mature plant can be grown.

A plant regeneration refers to the process in which one cell grows to a mature plant. A plant's regeneration ability differs from one plant to another, but by using today's tissue culturing technology, most of plants can be regenerated from the culture medium. But the regeneration process is still a black box and leaves rooms for further development, which in turn is the main point of transformation technology.

Regenerated plants go through the hardening process, which is an adaptation period, then gets planted into the soil. It is then cultivated and seeds are harvested. The whole process is described in Figure 3-7.

Crops transformed and regenerated go through the agricultural performance and utility test in order to determine which plants are appropriate to be grown into products. Most of the transformed cells

Table 3-1. Methods to Transform Plants with Recombinant DNA

Method	Explanation
Agrobacterium mediated transformation	*Agrobacterium tumefaciens*, with rearranged genes transformed, is used as a medium to grow and regenerate plant cells
Microprojectile bombardment	Recombinant DNA is coated on metal particles then is bombarded onto cells. This is the easiest way to use on a large number of crops.
Microinjection	Using a micro operator and a micropipette, the recombinant DNA is directly injected into the plant cell. It needs a lot of training and accuracy, which is the reason why the technique is limited
Viral vector transformation	A technique that uses recombinant virus to transform target cells. Very limited in the case of plants.
Protoplast transformation	With regenerable protoplast as the target, it utilizes chemical substance such as PEG with recombinant DNA, and thus is very limited usage.
Electroporation	With regenerable protoplast as the target, high voltage electric pulse is applied with recombinant DNA. Very limited usage
Liposome fusion	With regenerable protoplast as the target, the recombinant DNA is packaged in liposome.
Pollen tube method	Through the pollen tube made by fertilization, the recombinant DNA can be inserted. Very limited usage.
Dipping or imbibition	A method used on a few model plants such as *Arabidopsis thaliana*. When an organism is dipped in an Agrobacterium solution that includes detergents such as Silwet L-77, and then is cultivated, a transformed plant, without any tissue culture, can be obtained.
Microlaser method	Poke a small hole in the cell using a small laser and transform the rearranged DNA into the hole.

are selected out at this stage. The selected plants go through a more intensive risk assessment and breeding. Finally, biosafety information is handed in to the authorities, where the crops would be examined and approved to be commercialized on the market.

3-5. Marker Free Technology

As biotech crops being widely commercialized recently, there has been a concern that the selection marker genes used in the selection of genetic regeneration process might be spread to other plants or microorganisms. The possibility is very limited but the selection marker genes that give resistance against herbicides might move to other plants, and then move to the microorganisms in the internal organs of the animals and the people that consume the product. Thus, a transformation methods not to use selection marker gene was developed such as the method not to use the selection marker gene, methods to remove the selection marker gene after transformation, or methods to develop safer selection marker genes have been devised.

Cotransformation of the selection marker gene with desirable trait gene, and then removing the marker in the next generation was attempted. But in this case the problem was that the transformed traits were supposed to be made independently. The selection process by simple cross breeding was to be repeated, and it also had very low efficiency.

So recently, the selection marker genes are recombined within the *loxP* sequence, and then crossing with transformant with the *Cre gene* later leads to recombination in the loxP sequence. In this process, the selection marker genes in the middle would be removed.

Another method is using a marker safer than selection marker genes. This is a method that has been tried recently, using substances such as *phosphomanno-isomerase* (PMI) or fluorescent protein genes extracted from jellyfish in the place of selection marker genes.

In South Korea, prof. In Gyu Hwang's team in SNU is developing a technology to utilize the gene *toxoflavinlyase*, which degrades *toxoflavin*, a selective agent, and is being used in transformation of many plant organisms', even extending to microorganisms.

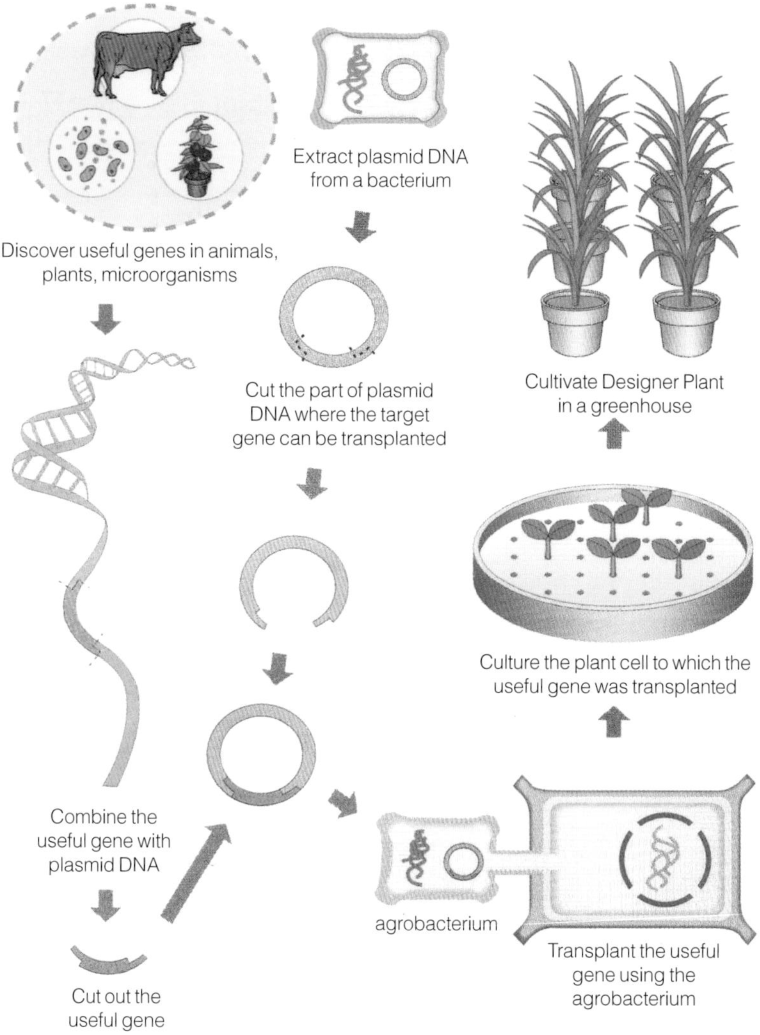

Figure 3-7. Gaining crops with transformed traits by transplanting foreign genes to a plant

3-6. Plastid Transformation

Most of biotech crop production is done by the usage of Agrobacterium or microprojectile bombardment to transform traits. However, these methods have met with some criticism. Low frequency of transformation through the integration of recombinant genes at a specific targeted site, unstability of the inserted recombinant genes, and the possibility of gene contamination into the ecosystem were pointed out. For example, when the herbicide-tolerant gene from the GM crops could spread to nearby plants and affect the weeds nearby. Superweeds could be produced. Otherwise, the gene contamination between plants could lead to the modification of native plants, and it could lead to the loss of the diversity in the gene pool.

However, these cases seem to be not as common as one can expect. Even in the traditional breeding utilizing artificial crossbreeding technology, the chance of cross fertilization is very low if the genetic backgrounds are different each other even in the same species. Let's assume that the pollen of a plant lands on the pistil of a plant of the same species but different genetic background. The pollen in this case, needs to get its sperm nucleus to the egg, the two needs to get fertilized, and they should not be degenerated in the process either. The chance to be fertilized even after going through all these steps is nearly impossible. Even though the fertilization gets complete and matures properly, there is much less a change the crop would germinate to grow into the next generation.

However in order to completely get rid of the low chance, scientists are developing a way to insert the gene in the cytoplasm, in order to make paternal inheritance impossible. A cell can be divided into the nucleus and the cytoplasm. Each organism has different chromosomes, and this is where the DNA, the code to life, is stored in. However, there is an organelle in the cytoplasm that contains DNA

Table 3-2. Comparison between Nuclear transformation and Plastid transformation

	Nuclear transformation	Plastid transformation
pros	-low in cost and labor	-Eco-friendly: overcoming DNA contamination through maternal inheritance -High expression level: 100~300 times that of nucleus insertion -Transgene silencing: none -Able to control insertion location and is not location-dependant either -Polycistronic expression possible -Blocks foreign proteins interaction: Blocks any unnecessary interaction with foreign proteins in the cytoplasm
cons	-Gene contamination -Relatively low expression level -Lower frequency of correct gene expression -inability to control insertion location makes expression level impossible to predict	-high costs, low efficiency
How to insert genes	-Utilization of Agrobacterium -microprojectile bombardment	- microprojectile bombardment
Development status quo	-common	- In development

as well: the mitochondria and the plastid organelle, the chloroplast.

DNA in the cytoplasm is of maternal inheritance, going from the mother to the offspring. Using these traits, a new trait-altering technology has been developed. By inserting the desirable trait gene into the genes of the chloroplast, the possibility of pollens from GM plants to fertilize and spread the foreign DNA to other species could be cut down. This is called "chloroplast transformation".

A plastid's DNA system is the same as that of a prokaryotic cell, which makes inserting in a specific site easier, and has the same high level of inserted DNA expression as in a bacterium.

A plant's plastid organelle is the place where the photosynthesis and many other metabolistic processes occur. By altering the DNA in the organelles, the altered metabolism process can increase crop productivity, and is able to grow high-value added crops. The plastid transformation process made it possible to stray away from the traditional way of determining traits, based on Mendel's genetic laws, and achieve the goal of GM crops of quality and quantity in an economical way.

Chapter Four

Biotech Crops in the Field

Now, let's look at the traits and breeding principles of biotech crops that have been or approved to be commercialized, and the developing biotech crops.

(In this chapter, characteristics and breeding principles of biotech crops are described by several examples of those approved as well as being developed.)

4-1. Examples and Breeding Principles of Biotech Crops

a. Herbicide Tolerant Crops

Farming is described as war against weeds. Yield loss due to weeds is estimated to be over 20% of the total yield and fertilizer is would be useless unless weed problem is solved. Thus, herbicide resistance would be the most desirable trait that biotech crops are equipped with. As explained in the methods of breeding biotech crops, herbicide tolerance genes could be used as selection markers for transformants harboring useful genes. Cells having a herbicide tolerance gene introduced together with genes of interest survive herbicide treatment proving that those useful genes are successfully introduced. In Figure 4-1, genetically engineered herbicide-tolerant soybeans grown in the field are compared with non-GM soybeans.

The breeding principles of herbicide-tolerant crops described in Figure 4-2 can be categorized into two parts. Herbicide bound to enzymes blocks their functions, and thus withers plants. For biotech crops categorized in the first part, genes mutated to encode target enzymes of herbicides structurally modified or genes from other organisms encoding the target enzymes are introduced to host so that herbicides are unable to bind and act on the target protein. For the second part, crops are transformed to express genes encoding enzymes to modify and inactivate herbicides. This can effectively in-

Figure 4-1. Biotech soybeans tolerant to herbicide
Left, Soybean field fully covered by weeds; Right, Soybean field after spraying herbicide

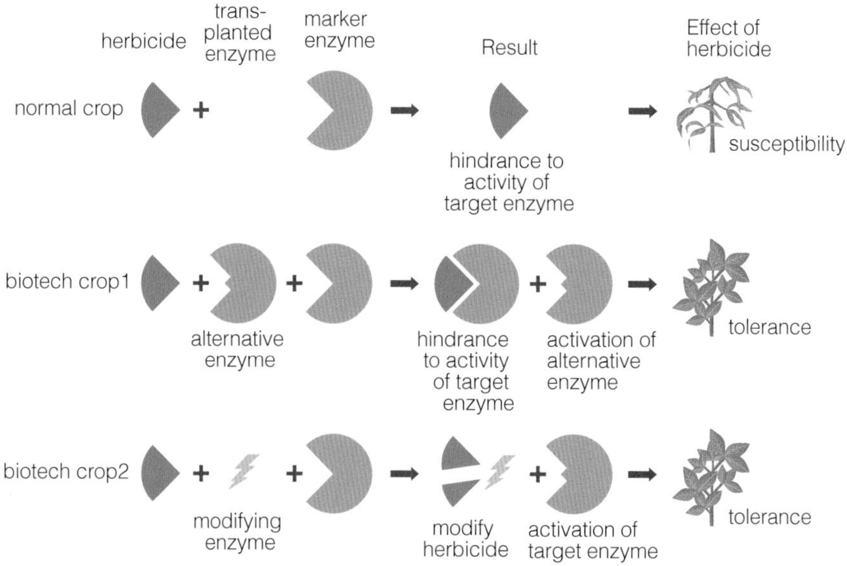

Figure 4-2. Principles of breeding crops tolerant to herbicide
Herbicide binds to a target enzyme and inhibit the activity of it so that plants wither to die. Bioengineered plants have either an enzyme with different structure, so that the herbicide can't bind, by expressing a gene from other organism or mutated (Biotech crop 1), or an enzyme inactivating herbicide (Biotech Crop 2).

activate the herbicide and lower the concentration of remaining herbicide in soil or crops, making biotech crops to be more eco-friendly and beneficial to human health.

Examples of crops bred by this principle are shown in Table 4-1. Companies producing herbicides came up with crops that can tolerate their herbicides. They are on one side blamed for selling the seed with herbicide, but on the other side it is really a thankful incident. When a gene that is resistant to a specific herbicide is transferred to weeds, farmers can use another herbicide to kill the weeds and grow crops that are still tolerant to the herbicide. The fact that growing crops that are tolerant to herbicide saves around 14% of herbicide usage proves the environmental friendliness of this technology.

In Korea, a research team led by Dr. Minhee Nam of the National Institute of Crop Science in Milyang bred a rice cultivar, Milyang 204, tolerant to herbicide Basta which is under risk assessment and expected to be commercialized (Figure 4-3).

b. Insect-Resistant Crops

An insect-killing protein gene of the microbe *Bacillus thuringiensis* that has long been used as an environment-friendly biological agent is used for breeding crops that are resistant to pests. Bacillus, a soil microbe, produces *Bt* protein (Bacillus toxin). This protein is dissolved inside alkaline stomachs of insects. Hydrolysis by digesting enzymes activates the protein which shows insect-killing effect by combining with a special receptor that exists on the insect's intestine cell membranes and then making holes on them. However, the receptor differs depending on the types of insects. Therefore, a *Bt* protein works specifically to an insect. Consequently, it can only be selectively used to narrow range of pests. The change in the structure of *Bt* protein determines the possibility of combining with the

Table 4-1. Examples of commercialized herbicide tolerant crops

Herbicide	Transgene	Principle	Target crop	Name of the product and the company
Glyphosate (Roundup®)	5-enolypyruvylshikimate-3-phosphate synthase (EPSPS) from A. *tumefaciens* CP4	Similar enzyme from microorganism	soybean, canola, cotton, corn	RRS Monsanto
Glyphosate (Roundup®)	*Ochrobactrum anthropi*'s glyphosate oxidase (GOX)	Oxidation of herbicide	canola, corn	GT73 Monsanto
Sulfonylurea (Glean®)	acetolactate synthase (ALS) from tobacco or Arabidopsis	Resistant mutant enzyme	cotton, canola, wheat, corn, carnation	19-51a Dupont pioneer
Phosphinothricin, glufosinate (Basta® Liberty®)	*Streptomyces*'s phosphinothricin acetyltransferase (bar, PAT)	Acetylation of herbicide	canola, bean, rice, corn	T25 Bayer (Aventis)
Bromoxynil (Buctril®)	*Klebsiella pneumoniae*'s bromoxynil nitrylase (BXN)	Hydrolysis of herbicide	canola, cigarette, cotton	BXN Bayer (Aventis)

Figure 4-3. Milyang 204, a herbicide tolerant rice cultivar bred for direct seeding
When Herbicide Basta is sprayed, non-GM rice plants wither to die are normal breed dry up to death but Milyang 204 which was reared through bioengineer technology is growing green.

receptor and the specificity toward insects, so it can be used selectively. Therefore, *Bt* protein does not affect untargeted insects, and is not harmful to humans and livestock whose gastric juice is acidic. In Canada, *Bt* pesticide is proven to be safe even to the extent of being sprayed by airplane to prevent pests in forest. Also, it is the only kind of pesticide that is allowed to be sprayed in organic farming (Figure 4-4).

Based on this fact, we can transform plants with a gene of *Bt* protein and let the plants produce *Bt* protein themselves, making insects to avoid the plants (Figure 4-5). Especially, pests like *Corn Ostrinia furnacalis* that cannot be treated with pesticides because such insects eat the inner part of the corn stem, can be more effectively controlled by this method than just spraying pesticides.

As the Table 4-2 shows, cottons and corns which are transformed with various genes encoding *Bt* proteins and resistant to specific pests have been developed and commercialized.

Figure 4-6 shows the difference in resistance of a corn transformed with a specific *Bt* gene against five different pests. *Bt*11 is effective against European *Ostrinia furnacalis* and has considerable amount of

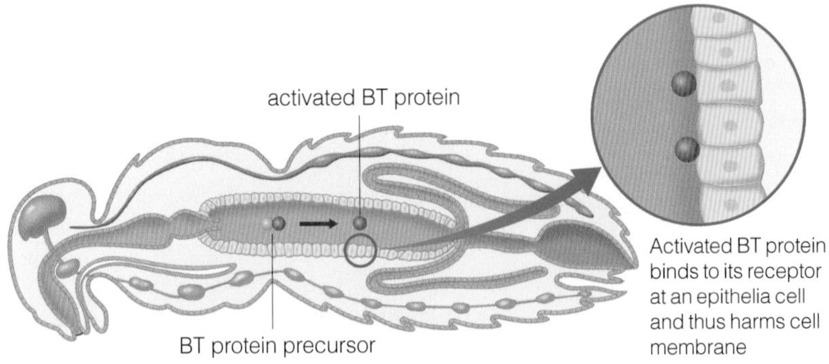

Figure 4-4. Working principle of BT protein
Human's stomach is acidic so BT protein activation does not take place and human's stomach does not have receptors so BT protein cannot work. Thus, it is harmless to humans.

Figure 4-5. Corn that endures pests
Left is the normal breed and right is the biologically engineered corn expressing BT gene.

Table 4-2. Pest resistant biologically engineered crops and selective pesticidal effect

Name of the item	Transplanted gene	Insect giving influence	Applied crop	Company
531 *(Bollgard)*	*Cry1Ac* gene (*Bt* ssp. *kurstaki*)	Lepidoptera	cotton	Monsanto
RBBT06	*Cry3A* gene (*Bt* ssp *tenebrionis*)	Adult Colorado potato beetle	potato	Monsanto
MON810 *(YieldGard)*	*Cry1Ab* gene (*Bt* ssp *kurstaki*)	European Ostrinia Furnacalis	corns	Monsanto
BT11	*Cry1Ab* gene (*Bt* ssp *kurstaki*)	European Ostrinia Furnacalis	corns	Syngenta
CBH-351 *(StarLink)*	*Cry9C* gene (*Bt* ssp *tolworthi*)	European Ostrinia Furnacalis	corns	Aventis
DBT418 *(BtXtra)*	*Cry1Ac* gene (*Bt* ssp *kurstaki*)	European Ostrinia Furnacalis	corns	Monsanto
1507	*Cry1F* gene (*Bt* ssp *aizawai*)	European Ostrinia Furnacalis	corns	Dupont
MIR162	*Vip3Aa20*	Lepidoptera	corns	Syngenta
MON863	*Cry3Bb1* gene (*Bt* ssp *kumamotoensis*)	Rootworm	corns	Monsanto

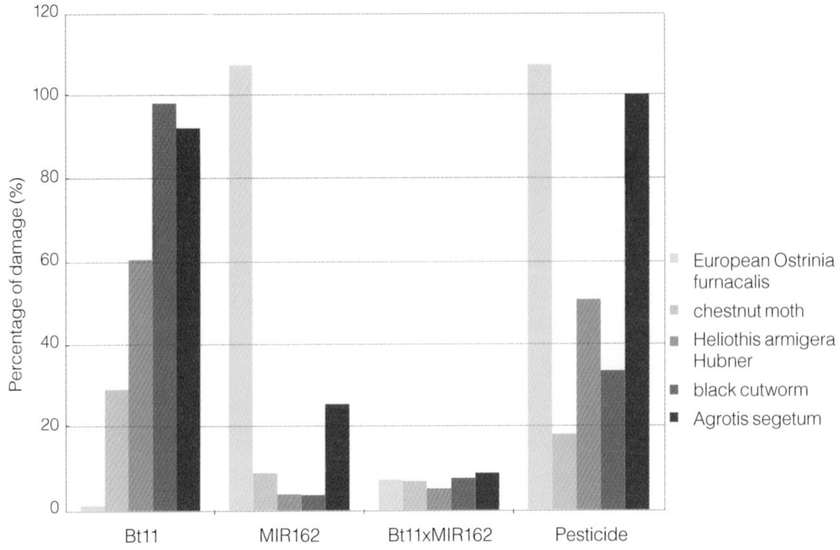

Figure 4-6. Comparison of damage by pests between biologically engineered crops and and pest control by pesticide.

Figure 4-7. Biotech rice that can endure Cnophalocrocis medinalis guenee
The normal breed on the left had been harmed but the pest resistant Bt rice with cry1Ac1 gene on the right did not damaged.

effect on chestnut moths, but is not that effective on other kinds of pests. MIR162 has no effect on European Ostrinia furnacalis. *Bt*11x-MIR162, which is the combination of the former two, has effects on most of the pests. These biotech corns showed excellent prevention effects compared to the chemical pesticide spray.

In Korea, Dr. Seok Cheul Seo's team at National Academy of Agricultural Science developed rice transformed with *m-cry1Ac1* gene, which was separated and mutated in Korea, that can stand against *Cnaphalocrocis medinalis* which causes tremendous annual harm. This is under the process of safety test and commercialization is anticipated.

c. Virus Disease Resistant Crops

Just like humans, plants get viral diseases and there is no medicine for both humans and plants. Because virus is smaller than bacteria and the cause of the disease is difficult to find, the importance of the prevention is highlighted more than ever. Animals and plants show immunity responses when foreign substances like viruses invading and, in the case of humans, we can prevent diseases by primary immune response activated by vaccine shots. Likewise, by introducing the virus shell protein gene or part of virus genome into plants and over expressing them, plants gain tolerance like the way people gain immunity by vaccine. The principle is known to come from gene's post-transcriptional gene slicing. Nevertheless, some drawbacks have been pointed out, such as the fact that this method results highly selective immunity and it can only be limited to very few viruses.

To see some examples that have been put into practice, papaya resistant to papaya ringspot virus, potato resistant to potato leafroll virus, potato virus X or Y, and pumpkin resistant to Cucumber mosaic virus (CMV), zucchini yellows mosaic and watermelon mosaic

Figure 4-8. Biotech crops that can endure viruses
Left top: in the picture, left is normal papaya and the right is biologically engineered papaya. Right top: pumpkins that were biologically engineered to endure viruses is more productive. Left bottom: tomatoes that were biologically engineered to endure viruses is more productive. Right bottom: Potatoes that were biologically engineered to endure viruses is more productive.

virus can be listed. In Korea, especially the team of Dr. Ji Hak Han at Nongwoobio company has developed red pepper that is tolerant to CMV and it is under the process of safety tests so we are waiting for the commercialization.

d. Oil Crops with Improved Fatty Acid Composition

Since animals cannot produce unsaturated fatty acids by themselves, they need to gain it through foods. However, poly-unsaturated fatty acids easily acidify in the air and produce peroxides, which smells and is bad for the health. To improve this, we control the degree of unsaturation through chemical hydrogenation and trans-fatty acid is produced in this process. In the case of oil plants that

Figure 4-9. Red pepper that endures viruses
The right ones do not grow well due to CMV but the left biologically engineered red peppers are growing well.

produce lipid, through bioengineering technology, fatty acid can be controlled so lipid can be improved without transfat.

By transforming with a gene encoding *12:0-ACP thioesterase*, which is involved in synthesizing short fatty acid, laurate (C12:0) and myristate (C14:0) which are included in shampoo, detergent, etc were increased. Also, by over-expressing the gene of stearoyl ACP desaturase, which unsaturates fatty acids, the amount of unsaturated fatty acids was increased. In reverse, a new cooking oil with increased amount of the mono-unsaturated oleic acid and decreased amount of poly-unsaturated linolenic acid by silencing its own fatty acid unsaturated enzyme to lengthen the distribution period without trans fatty acid is waiting for marketing.

ω-3 fatty acid* is known to have good effects on growing children's brain development so many nutritionists recommend eating it and consumers mostly like it. However, this ω-3 fatty acid is usually contained in external blue colored fish such as mackerel so consumers and children who do not like fish are difficult to consume. Thus, if we separate gene of enzyme that works to synthesize ω-3 fatty acid from fish and transplant it in plants, we can gain fatty acid ameliorated soybeans oil, rapeseed flower oil and many other plant oils.

In Korea, there are cases where the fatty acid composition was changed, storability was increased and where amount of tocopherol was increased to fortify the essential vitamin and to improve its storability.

Table 4-3. Biotech oil crops whose fatty acid component was improved

Transplanted gene	Transformation trait	Applied crop	Company of invention
stearoyl ACP desaturase	high-unsaturated	canola	Pioneer Hi-Bred
12:0-ACP thioesterase	high laurate (12:0) and myristate (14:0)	canola	Calgene
anti-sense fatty acid desaturase (GmFad2-1)	high-oleate (monovalent unsaturated fatty acid) low-linoleate (multivalent unsaturated fatty acid)	bean	DuPont

* ω-3 is unsaturated fatty acid and usually is in blue back fish, salmon and seals. The main component is DHA (Docosahexaenoic Acid) and EPA (Eicosapentaenoic Acid).

e. Plant with Improved Starch Contents and Structures

When ADP-glucose pyrophosphorylase gene from E coli, which is involved in glycogen synthesis was expressed in potato, starch content increased 30 to 60%. Therefore, if we fry these potatoes in oil, these potatoes absorb less oil, making them to be crunchier. On the other hand, if this gene is introduced in reverse direction, the expression of the gene and the enzyme are decreased so the sweetness increases. There was also a case where a gene encoding wheat puroindoline which function to bind to starch granule was expressed in rice to seed starch viscoelasticity softer. In this way, characteristics of starch can be changed by the over or under expression of genes encoding enzymes involved in starch and carbohydrate biosynthesis.

For the purpose of effective bio energy production, by the over-expression of thermo-tolerant α-amylase gene, there is no need to add any α-amylase for alcohol fermentation.

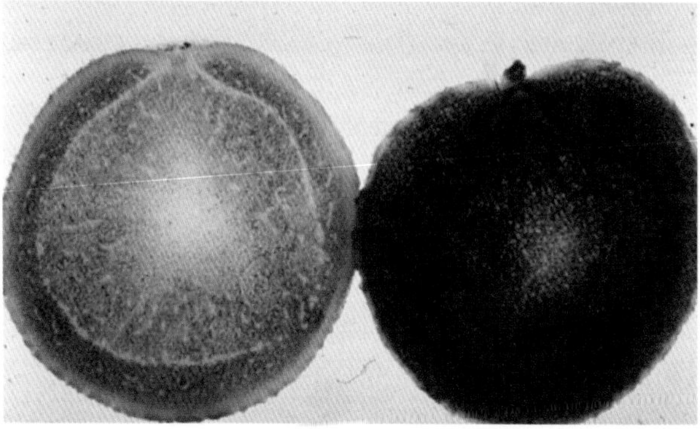

Figure 4-10. Potatoes with increased amount of starch
A potato (right) over-expressing ADP-glucose pyrophosphorylase gene had increased in the amount of starch it contained so iodine response became much darker which was dark blue.

f. Delayed-Ripening Tomato and Biotech Vaccines

For developing delayed-ripening tomato Flavr Savr, the biotech tomato commercialized for the first time, transformation was not done by the insertion of an external gene but by the introduction of its own gene segment encoding polygalacturonase in the antisense direction. By doing this process, it restricts the expression of the genes that hydrolyze pectin and prevents the softening of cell walls so people could pick and transfer tomatoes after they were ripened enough, causing tomatoes to have high solid content. Scientists are in the process of developing tomatoes with more flavor and texture through the restriction of *1-aminocyclopropane-1-carboxylic acid synthase* gene or through the increased activity of 1-aminocyclopropane-1-carboxylic acid deaminase, causing ACC concentration to decrease, ultimately restricting the synthesis of aging hormone ethylene.

Other than this, tomatoes with more lycopene, as proved its effectiveness in Golden Rice, which is the precursor of vitamin A, anticancer tomatoes with drastically increased amount of antocyanin by transformation with two transcription factor genes, Delila and

Figure 4-11. Compared to normal tomatoes ketchup (left) that separates with liquid easily, ketchup made with not softening tomatoes has more solid content so it is more thick and delicious.

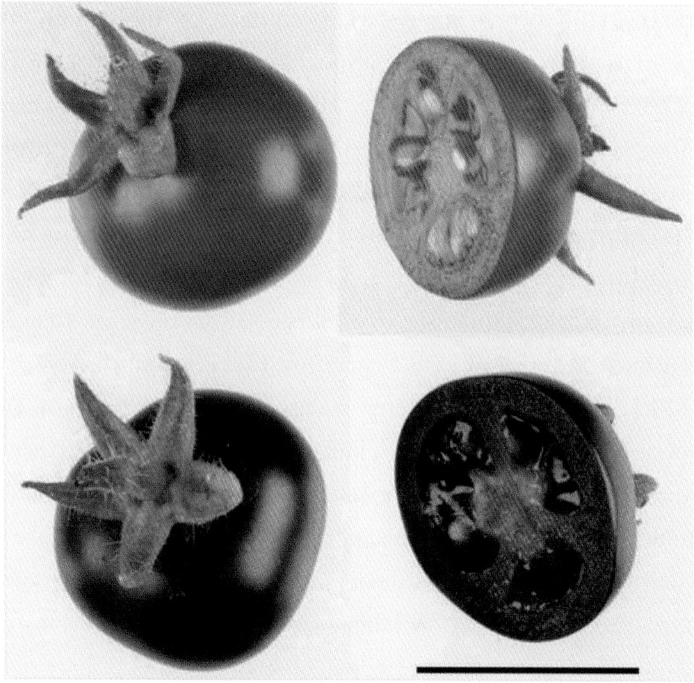

Figure 4-12. anticancer tomatoes that produce anthocyan
There is no anthocyan in the normal tomatoes on the top but there are 2.83±0.46mg of anthocyan in the biologically engineered tomatoes on the bottom. Nature Biotech 26: 1301 (2008)

Rosea 1 from snapdragons, and tomatoes that produce vaccine, for which we can take an advantage of consuming them without cooking is expected to be commercialized.

g. Biotech Crops Tolerant to Environmental Stresses

Basically, crops are sensitive to environmental stresses such as drought, salt and low temperature so these stresses lead to less productivity. However, plants have self defense genes that respond to environmental stresses so plants survive and adapt to changed environment by expressing these genes. Although we are living in a world of abundance owing to stress tolerant crops traditionally cross-bred, extreme weathers due to global warming take agriculture to the extreme situation. Therefore, studies to develop crops tolerant to environmental stresses are underway by controlling genes involved in stress resistance based on molecular breeding technology. There are cases in that genes for producing substances protecting plants from stress such as proline, betaine, mannitol, trehalose, tocopherol, flavin and antioxidants are over expressed to increase tolerance against stress and that genes encoding transcription factors, such as ABF1, CBF1, DREB1A, etc. to regulate the expression of these genes or genes encoding protein kinase to modify signaling pathway. In addition, transgenic plants tolerant to toxic heavy metals, such as cadmium, mercury and copper, which restrict crop growth are under development. Benefits of these plants are effective phytoremediation of soils contaminated with heavy metals.

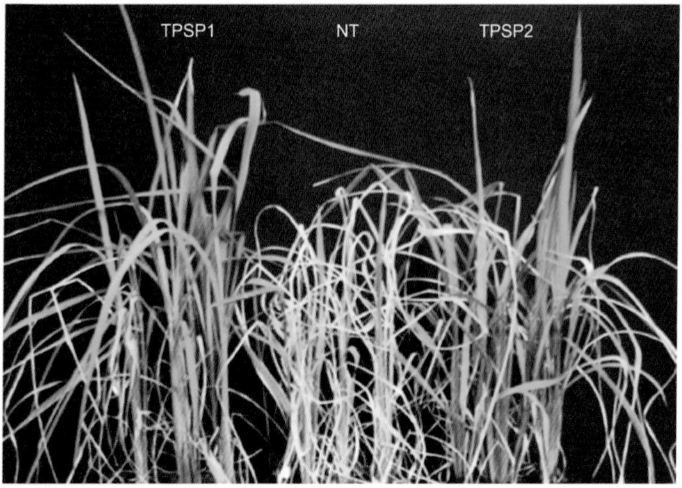

Figure 4-13. Super rice that endures drought
It is normal rice in the middle and the rice on the left and right are rice with *trehalose synthesis* gene transplanted.

The team of Dr. Joo Gon Kim at Myoungji University invented rice that can endure drought by using *trehalose synthesis* gene. The result was reported by starting from England's BBC, then America's CNN, CBS and so on as Asia's hope and, in Korea, it was introduced as super rice that even grows in deserts. Korea played a leading role in relocating this technology which was invented by cooperating with Professor Ray Woo at Cornell University in USA to the Indian bioengineering company, Mahyco for commercialization.

h. Plants Resistant to Bacterial and Fungal Diseases

Agriculture has integrated crop farming systems and changed to single crop farming systems to increase productivity. Therefore, it became easier for pathogens to proliferate and plant disease became more prevalent. To find solution for this, there have been studies on resistance to disease in plants together with one to develop disease management techniques underway.

Since plants are not locomotive, they are equipped with various physiological functions to protect themselves from insect or microorganism infection. Depending on plant cultivars, plants show various resistance responses against pathogenic microorganism. Resistance responses of plant cell to pathogenic microorganism are categorized in two groups in that the first group includes the local response restricted to the area of infection and the second group is characterized as systemic acquired resistance in which the response is expanded to the entire plant body. Local response can be seen at the infected part or the part around it. In this case, synthesis of antibiotic substances that restricts the growth of pathogenic organism or hypersensitive response which is the death of few cells to prevent the expansion of pathogenic organism occurs. These resistance responses can be seen when a plant is infected by viruses, molds, and bacteria. Many Rgenes which induce these hypersensitive responses were identified. *β-glucanase* and *chitinase* genes were expressed or hypersensitive reaction caused by ribosome inactivating protein was used to increase the disease resistance. With the development of crops transformed with these genes, there are studies going on about inducible promoter to express resistance genes only when the plants are infected by pathogenic organism.

Salicylic acid and jasmonic acid are known as the signal transducers which can induce resistance responses regardless of the variety of plants or pathogenic organism. Salicylic acid is a signal transducer which induces the expression of pathogenesis related protein around plant's uninfected parts and ultimately spreads acquired immunity all over the plants for second infection. And jasmonic acid is a signal transducer which makes plant to gain resistance against some pathogenic organisms and wound causing pests. In addition, when pests make wounds on plants, jasmonic acid is synthesized all over

Figure 4-14. Biotech rice that endures grain bacterial grain rot Enzyme gene that dismantle B. glumae's toxoflavin that causes bacterial grain rot is transplanted in the rice, left is the normal rice.

the plants so that even uninfected parts gain resistance by synthesizing protease inhibitor beforehand. Moreover, jasmonic acid also has properties of plant hormone which is involved in the development of seeds, aging, photosynthesis, and other developmental processes. Recently, disease resistance of Arabidopsis transformed with *jasmonic methylation enzyme* gene has been observed.

i. Ultra Productive Rice with Immensely Increased Yield Productivity

There are many ways to increase crop yield. An annual yield increases when weeds and pests are efficiently controlled and when environmental stresses such as drought and low temperature are well managed. However, compared to these indirect ways of increasing production by reducing the losses, a direct way of it by increasing the

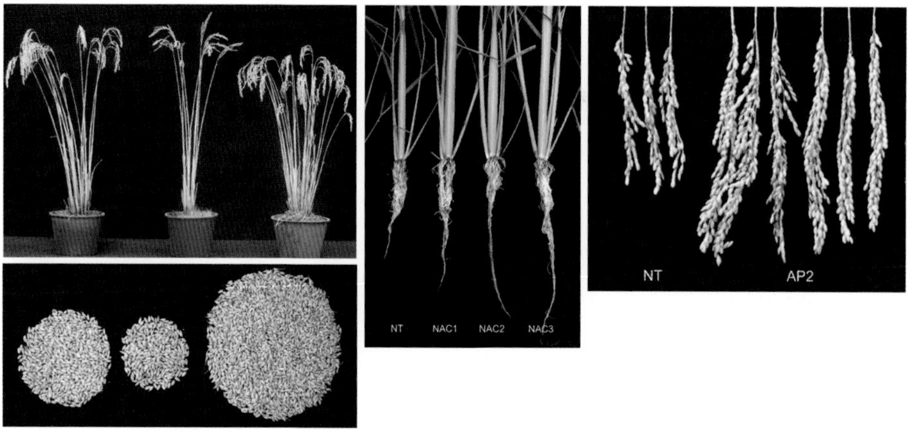

Figure 4-15. Ultra extremely productive rice that hugely increased the yielding of rice
Left picture: rice in the middle which has destroyed NAM gene compared to normal rice has decreased amount of rice but rice with over expressed NAM gene on the right is extremely productive.
Middle picture: Thanks to transplanted NAC gene, the stem of rich gets thicker and becomes more productive.
Right picture: Transplanting and over expressing AP2 gene makes rice to gain bigger ears, leading to more productivity.

rate of metabolism of plants can be considered as the fundamental solution for increasing need for food. Crop functional genomics center has developed rice with immensely increased yield productivity through collaboration of researchers.

By using some transcription factor genes (Ann, NAC, AP2, etc) that control the expression of genes, productivity drastically increased and it was clear that such increases are caused by the effect of increased metabolic rate such as bigger sizes, more seeds due to increased tillering and heavier weight due to thicker stem.

j. Coffee without Caffeine

One enjoys caffeine sometimes but has to restrain coffee due to caffeine in certain occasions. For such occasions, coffee without caffeine is under research. The coffee in market is not coffee without caffeine but decaffeinated coffee, made by extracting caffeine from normal coffee. Decaffeinated coffee is less preferable because the flavor and aroma chemicals are also extracted during the decaffeination process. However, the biotechnology allows coffee to remove 70% of the caffeine and preserve other flavor and aroma chemicals by suppressing theobromine synthase gene during the multi-staged biosynthesis process. The situation is very much the same with teas.

By applying the similar technology, poppies without morphine can be developed. Poppy oil is the only cooking oil that can be prescribed to heart patients. However, poppy cultivation is strictly prohibited because morphine component that is unique to poppies is classified as narcotics, thereby poppy oil is hard to find. By using the same technology that is used to restrain biosynthesis of caffeine in coffee, people can obtain morphineless poppy when one gene of codeinone reductase, morphine biosynthesizing enzyme, is suppressed. This is good news not only to heart patients but also to general common people who are able to appreciate beautiful poppy flowers freely. Cigarettes without nicotine made using this technology is currently in market.

k. Blue Roses and Carnations

"Blue rose" is an expression for "impossible". However, a blue rose is possible with biotechnology. Florigen, an Austrailian biotechnology company, obtained a blue rose whose petals' anthocyanin is 100% composed of blue pigment, delphinidin, by suppressing a dihydroflavonol reductase gene that biosynthesizes red cyanidin and orange pelargonidin pigments. Dihydroflavonol reductase gene was isolated from iris and overexprressed it to synthesize blue pigment delphinidin. Although the rose is not perfectly blue as expected, but as research goes on, we may be able to enjoy fantastically blue rose that could not be seen in nature in near future. Also, by controlling a gene encoding an enzyme working at the step of dihydrokaempferol metabolism, which is a branching point of biosynthesis of anthocyanin, blue carnations with different colors have been commercialized (Figure 4-16).

We can also obtain blue cottons by this method. Blue cottons will bring about innovative changes to textile and fashion industries by making dying process unnecessary. Because dying process severely causes pollution, if the process is omitted, it is obvious that environ-

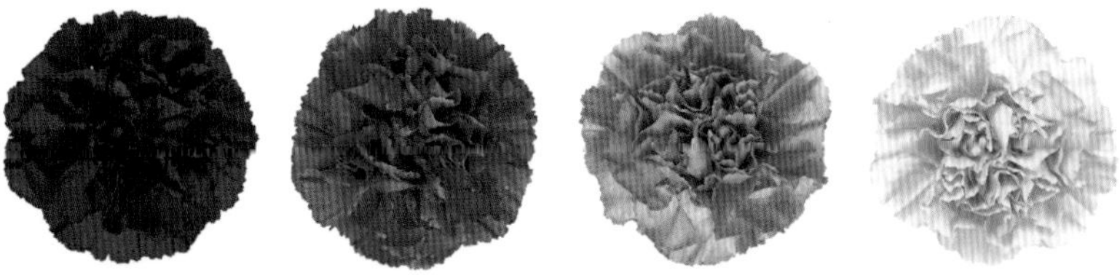

Figure 4-16. Australian company Florigene's blue carnations, Moonvista™, Moonshade™, Moonlite™, Moonaqua™, which had been developed and marketed.

Figure 4-17. Blue roses developed by Australian company Floregene and Japanese company Suntory.

ment-friendly cotton products including blue jeans would get the spotlights.

1. Landmine Detecting Plants

Many children in countries that are experiencing war are being suffered from landmines because countries are less interested in getting rid of landmines due to the higher cost for removing. High removing cost is mainly due to difficulties to find landmines buried under ground. A Danish biotechnology company invented a plant that changes its color by biosynthesizing anthocyanin pigments when explosive chemicals are around. If an annual plants that lacks the ability to make offspring are air-spayed, landmines-buried regions will be easily detected and such plants will vanish next year thanks to its impotency. These plants make us think that developing new type of plants could be limited because of the lack of human imagination required to conceptualize products and not because of technological

Figure 4-18. A plant that produces anthocyanin and turns red by chemicals from landmines

limits. Therefore, finding out the gene that backs up the concept of products is said to be both the beginning and the end of the job.

m. Duplicate Gene Transplant and Gene Stacked Event

Efficient pest, weed, and disease controls are most important features, next to high yield, for agriculture. Since the aforementioned herbicide tolerance and pest resistance are the most basic requirements for biotech crops, all future biotech crop developments will start from transforming genes with such traits. In 2007, biotechnology crops having both of such traits were already 19% of total biotech crops. To breed crops with multiple traits, we either introduce multiple genes of desirable traits at the same time (multiple gene transfer), or stack genes by cross-breeding between lines harboring single genes transformed (gene stacked event). These biotech crops with multiple genes will be prevalent in the future. Currently, Smartstax®, harboring six pest resistance genes and two herbicide tolerance genes, is waiting to be released.

4-2. Current State and Prospect of Biotech Rice Research

Rice is a staple crop for 3.8 billion people and is produced and consumed mostly in Asia. For past two to three decades, increase in rice production was mostly owing to labor-intensive producing technology and the development of traditional rice-breeding technology. Consequently, the present rice production has nearly doubled compared to that of 1965.

However, nutritional unbalance in developing countries is caused by deficiency of main nutrients and minerals mainly due to the consumption of only few staples (rice, cassava, wheat, and corns). And rice causes malnutrition from nutritional unbalance because it lacks nutrients like protein, vitamin, and minerals. Therefore, developing countries that only eat rice have serious starvation and disease problems. In the World, there are half million children who become blind due to vitamin A deficiency and 1 million 150 thousand children are dying from the same reason. Especially, in Southeastern Asia, 70% of children under the age of 5 are suffering from vitamin A deficiency, and diseases caused by vitamin A deficiency are increasing in number. Moreover, in the world, 2 billion people are suffering from iron or iodine deficiency.

Golden Rice, a product of rice research to overcome nutritional deficiency, is developed to fortify vitamin A that is deficient in rice by increasing the content of its precursor, β-carotene, and has just

normal rice Golden Rice Golden Rice 2

Figure 4-19. Golden Rice: it becomes the carrot color when β-carotene, a precursor of vitamin A, increases.

started to be bred in the Philippines recently. UNICEF is expecting to save 1 million to 2 million children from death annually through the development of vitamin A fortified rice.

From about 10 years ago, the level of plant transformation technology and application based on molecular breeding has been improved to the extent of developing types of rice that resists drought, pests, diseases, and is extremely high-yielding; even rice that contains quasi-insulin growth factors for diabetics is in development (Table 4-4). However, there are no transgenics commercialized, while Bt rice that resists pests is waiting for permission for commercial cultivation.

Rice has been thoroughly studied as a model for plant molecular biology, since its genome size is relatively small, which is 4.3x108bp, and was recorded as the first living organism whose DNA sequence of genome is analyzed 4 times already thanks to hard work to examine the structure of the particular gene. Until now, because improvement of plants by traditional breeding technology has constantly been studied and tremendous genetic knowledge has been amassed,

Table 4-4. Rice transformants and transplanted genes

Purpose of transformation	Transplanted genes
Pest resistant rice	*Bt (crylAb)*
RSV resistant rice	*coat protein*
Leaf-blight resistant rice	*toxoflavin lyase*
Herbicide (glufosinate) resistant rice	*PAT (bar)*
Drought resistant rice	*trehalose synthase*
Extremely high-yield rice	*NAC, Ann,* or *NAM*
Vitamin A (β-carotene) reinforced rice	*phytoene desaturase* *carotene desaturase*
Diabetes rice	*human insulin-like growth factor 1* or *Glucagon-like peptide-1*
Low-protein rice for brewing	*antisense-glutelin*
Low-allergen rice	*antisense-albumin*

genetic information from completely decoded genome will efficiently enable the improvement of rice cultivars. Moreover, rice transformation techniques with useful genes becomes widely spread to be a revolutionary basis for transgenesis and will be a foundation to solve 21st century food problems.

4-3. The Status of Korea

In Korea, even though lab-level biotechnology crop development cases are numerously reported, there are no commercialized products yet. For the last 10 years, through various research and development projects were supported not only by the Ministry of Science and Technology's new-functional biological material development project, but also by the Rural Development Administration, Korean Science and Engineering Foundation, and National Research Foundation of Korea. Rural Development Administration (14 crops and 35 species), Korea Research Institute of Bioscience & Biotechnology, and almost all university labs in the country are developing herbicide, diseases, and disasters-resistant, or functionally reinforced biotech crops such as rice, tomatoes, potatoes, beans, red peppers, Chinese cabbages, garlic, perillas, cigarettes, petunias, chrysanthemums, and roses (Table 4-5). Through researches, herbicide-tolerant rice and potatoes made by designing these plants to deactivate the herbicide glufosinate (Basta®) by introducing bar gene that encodes phosphinothricin acetyltransferase (PAT) were reported. Besides, perillas that contain increased amount of ω-3 fatty acid with δ-15-desaturase genes, environmental stress tolerant and antiaging cucumbers made by using superoxide dismutase genes, disease-resistant tobacco made by using jasmonic acid methyl transferase genes, desiccation-resistant tobacco made by controlling abscisic acid responsive factors are being developed and they currently are reported to academic communities or under patent-pending. Scigen-Harvest, a biotechnology company, obtained transformants with 20% increased number of tillers and panicles by transforming rice with Protox (proto-porphyrinogenoxidase), a herbicide-tolerant gene, and is performing assessments at large-scale in the field. Also, Rural Development Administration crossed herbicide-resistant transformants with Milyang-23 rice and selected lines with normal fertility; the Institute reported that some of them showed increased yield productivity when crossed with Tongil-type Samgang rice (1,33kg/10a). If these transformants are used as a parental line to breed hybrid rice, drastic increase in yield will be expected. Recently, Professor JuKon Kim's team at Myongji University developed drought-tolerant trehalose rice and extremely high-yielding rice and licensed these techniques to India and Germany, respectively.

Development of agricultural biotechnology in Korea is lagged behind world trends mainly because of funding systems in which small projects with short-term objectives are majority of projects, lack of uniqueness in research due to the absence of basic research that requires long-term and continuous investment, and no active attitude of the government due to repulsed consumers. However, recently, results seem to become tangible shortly because Koreans try hard to secure uniqueness of research results of Korean main crops such as ginsengs, red peppers, Chinese cabbages, garlic, watermelons, etc. Especially, at a time when controversy over of transgenic food and environmental safety becom intensified, it is very encouraging that people also focus on developing safer transgenic technology. Although the commercialization was lagging behind also because of incomplete legal systems, it is expected that biotech crops will be pursued actively soon mainly by the Rural Development Administration since standards and methods for food and environmental safety assessment was established in 2001.

Especially, technology regarding traditional crop breeding and cultivation that is essential for the commercialization of biotech crops is a world-class level. e Korea has accomplished a green revolution that enabled self-sufficiency of rice in 1970s. If Koreans achieve commercialization through reinforcing classical breeding by relating it with new gene transformation techniques, there can be a synergy effect through fusion of traditional and new technology.

World seed market has reached $21 billion at 2000 and expected to grow to 30 billion dollars by 2010. Seed market size of biotech crops used to be only 3 billion dollars in 2000, but people are expecting that the market will grow to 25 billion dollars by 2010. This expectation insinuates that high percentage of crops (about 80%) will be substituted by bioengineered crops.

However, current Korean vegetable seed market is only about 150 billion won (one thousand million dollars), which is 0.5% of the world market. To follow the world tendency, the market is expected

Table 4-5. Bioengineered crops that are being developed in Korea

Feature	Crops
Herbicide-tolerant	Rice, red pepper, potato, Chinese cabbage, cabbage, cucumber, perilla, garlic, watermelon, bean
Disease and pest-resistant	Rice, red pepper, wheat, Chinese cabbage, potato, cabbage
Increase in productivity	Rice with increased photosynthesis efficiency
Environmental disease-resistant	Drought resistant rice, salt·drought·heat resistant potato
Increased quality	Perilla with improved fatty acid
Special function	Vaccine producing tomato, blood pressure decreasing perilla, vitamin E reinforced perilla and lettuce
Value added improvement	Flower-color changing lily, early-blooming chrysanthemum, bloom-controlling red pepper, apple

to grow to 200 billion won, but if there is no technological self-development, Korea should only depend all on imports.

4-4. Effects of Bioengineered Crops

In the past, agricultural yield productivity has been achieved by using pesticides to decrease the damage of pests and diseases and fertilizers to increase productivity of the soil, respectively, but bioengineered crops are now solving most of farmers' chronic problem of weeds, pests, and diseases and are causing high soil productivity. Figure 4-20 shows biotech crops' effects on reducing herbicides and pesticides (Brookes and Barfoot, 2008). Amounts of herbicides and pesticides are reduced by 7.9%, but if they are calculated applying environment load quotient considering the effect on environment, 15.4% is decreased. This fact manifests that GM crops are rather environment-friendly.

Table 4-6 shows the bioengineered crops' increased yield effects over the last 10 years (Brookes and Barfoot, 2008). The increased

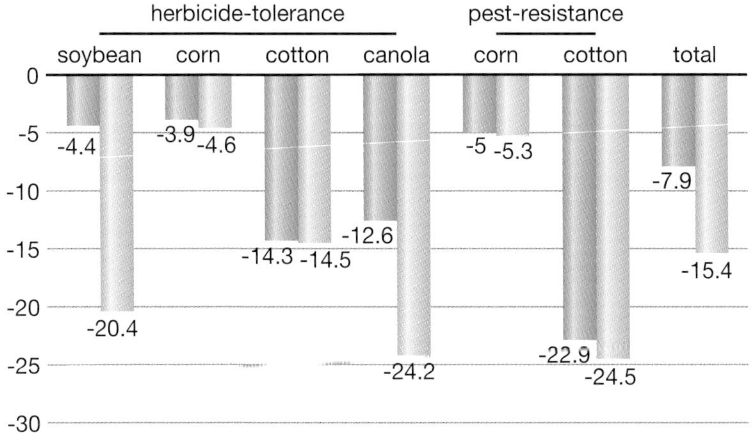

Figure 4-20. Effect of GM crops to decrease herbicide and insecticide use (1996-2006)

Table 4-6. Increasing yield effects of bioengineered crops

Crop	Bean	Corn	Cotton	Canola
Total increase (million ton) in 1996-2006	53.3	47.1	4.9	3.2
Increase in 2006* (million ton)	11.6	9.7	1.4	0.2
Increase percentage in 2006 (%)	20	7	15	3

* Increased amount in 2006 equals food of 67 million people
** Brookes and Barfoot, 2008

amount in 2006 is equivalent to the amount of food that can feed 67 million people. The effects will be more noticeable as time passes because the next generation's yield-improved bioengineered crops are developed on the basis of herbicide-tolerance and pest-resistance that are currently being developed.

Since the world population is exploding, developing biotech crops is rising as another agricultural revolution after the Green Revolution. Contemporary agricultural bioengineering technology not only contributes to increased productivity, environmental conservation, and improvement of safety and quality of food, but also is perceived as a strong alternative that can enhance agricultural competitiveness. Although developed countries like the United States and Canada are leading the development of bioengineered farm produces, Korea is not so much falling behind compared to such countries in terms of technology. Thus, biotechnology is expected to be accepted as a promising technology. Koreans should not give up various advantages of bioengineered farm products considering the Korea's dependence of most farm products on imports and low self-sufficiency rate.

Chapter Five

Risk Assessment and Regulation of Biotech Crops and Foods

Recently, the research and development of Biotech crops are being carried out actively and its worldwide commercialization is also progressing rapidly. Biotech crops have been a food resource for more than 45 hundred million people of 52 nations for over 10 years and a major feed for livestock. There is no evidence that GMO crops are not safe, Despite of their economic and environmental benefits, some consumers worry about the potential risks of Biotech crops to the human and environment.

Current worries of consumers are not much different from the questions that were raised when biotech crops were first commercialized in 1994. Most of those questions have been resolved as Biotech crops were grown and consumed on a large-scale for more than 10 years. Potential risks raised by experts are the following: dangers of new toxic substances, allergic reaction, change of essential nutrients, antibiotics tolerance, and threat to human bodies from long-term intake of Biotech crops.

Furthermore, concerns over the possibilities of destructing biodiversity and the decrease in number of beneficial insects due to the disturbed environmental and ecosystem, emergence of insects with tolerance to insecticides, transformation of Biotech crops to weeds, genetic contamination caused by transfer of modified genes to other plants, and creation of super weeds were raised. Not only such unintentional negative effects are worrisome, but whether it is ethical and morally tolerable to modify living organisms is also being controversial issue. Also, experts pointed out that future social and economic influences resulting from oligopoly or monopoly of Biotech crops' production and development by multinational corporations may result in subordination of agriculture and economy of underdeveloped countries.

These biosafety controversies over Biotech crops became hot issues when British Dr. Arpad Pusztai publicized his animal experiment results. When he compared mice fed on GMO potatoes in which *lectin* gene was transformed with mice fed on wildtype potatoes, more deformities were detected in the small intestine of the mice fed on GMO potatoes. However, the modified potatoes used in the experiment were products only to be used for research with no intent to commercialize, and Dr. Pusztai's experiment could not be reproduced nor provide enough evidence to support statistical significance (refer to example in Chapter 6).

Factors such as negative report of news media, non-scientific misunderstandings of diverse civic groups (Figure 5-1), suspicion of the industries and exaggerated advertisement for commercial purpose, and lack of knowledge on agriculture all worked together to initiate citizens' campaigns against Biotech crops blindly. Either scientists did not appropriately respond to consumers' worries about Biotech crops and thus failed to persuade them about the necessity, value, and effectiveness of this new technology.

Despite such a negative movement from civil groups, for over 10

Figure 5-1. Citizens' campaign demonstrating against biotechnological crops

Figure 5-2. NON-GMO declaration from food companies

years Biotech crops have been cultivated in more than 8% of the entire farmland. GMO cultivation is annually increasing by 13% in 23 nations. The potential threats raised in the earlier stages are being resolved as the possibility that they may happen is unlikely.

Science and technology is double-sided with both risk and convenience. It is just like as we cannot give up the convenience of car even though it emits exhausts and has a risk of accident. In fact, the technology itself is neither good nor evil. Good and evil are determined by how people use it. Of course unintended problems beyond our imagination may occur. However, just prohibition or restriction of all research would not solve the problem. Consumers must have more interest in the new technology and try to learn more about it in order to make sound judgments. Moral and social problems and issues raised from fear can be solved only when we understand the fundamental principle of science.

The revolution in biotechnology will influence more widely in both size and scope than any other technological revolution even to the cultural and social extent. Designer Plant technology in particular will be established as a survival technique for the human race. Similar to the Designer Baby technology, it will definitely lead the bioindustry and prove the potential of the technology and industry.

Biotech crops are not allowed to commercialize immediately following development. From the conceptualization of product to the search for the gene to utilize, safety is considered to be the top priority, and prototypes are very strictly tested in performance demonstration experiments. Most products fail in this step, and only few of which effectiveness and safety are proven become commercialized. In such process, researchers must take systematic and complete risk assessment tests, then organize the data to apply for approval. Developers usually make decisions in early process so that they can minimize the loss if their products get disapproved.

In most countries, genetically modified agricultural products must take environment and food risk assessment tests, and only when they

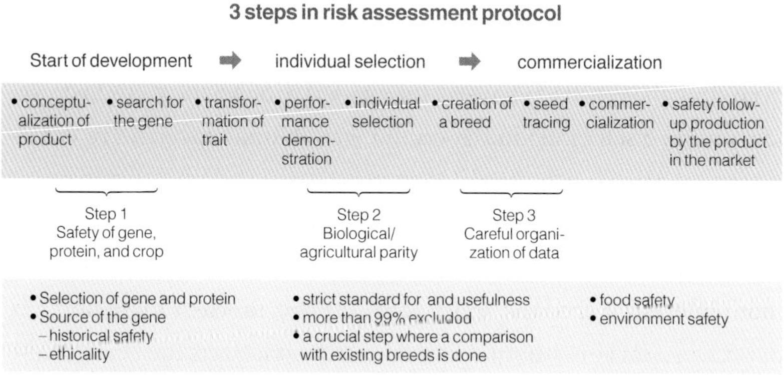

Figure 5-3. Risk assessment protocol that biotechnological crops should mandatorily go through before commercialization

pass the tests can they be used. It is similar to how newly synthesized drugs or pesticides are approved as commercialized products only after the strict risk assessment tests. But beforehand, the companies which developed the products also carry out risk assessment tests of their own.

When researchers develop GMO products, they consider safety most important in planning and carrying out experiments. If the results show any sign of risk to the human body or environment, then they must stop development.

For example, an American seed company Pioneer Hybrid had extracted 2S *albumin* gene, especially rich in methionine and cysteine, from Brazilian nuts and transformed it in soybeans in order to increase the sulfur-containg amino acid content in soybeans. However, allergic reactions to Brazilian nut had been often reported, and a new protein synthesized by transformation was found to be the cause. In a self risk assessment test by the development company, the problem that genetically modified soybeans can also cause allergic reaction was raised, and thus the product was discarded. This case shows how developers carry out risk assessment tests for themselves.

Another example is Starlink corn debacle in 2001. The news that some of insect-resistant corn that were approved as feed was used to make taco shells had shocked the world. Aventis Crop Science, the company that supplied such corn seeds to make taco shells, voluntarily cancelled out product registration, recalled all the remaining distributed corns, and said it would compensate financially when necessary. Aventis Crop Science lost nearly 1 billion dollars, which was not even close to the anticipated revenue. As this example shows, GMO products cannot be rashly commercialized. because of the great loss that follows when developing products have problems.

Unlike the original intention to harvest cleaner and safer agricultural products, Biotech crops raised questions over whether they are safe to the human body and environment, and the production and consumption of Biotech crops became a controversial issue. In response, governments set up the systems to secure safety objectively. Figure 5-4 presents the risk assessment system and specifications of Biotech crops. Details will be discussed later, but what should be noted is that the system is not just for the approval but for continuous safety follow-up by tracing the product after its commercialization in the market.

As the benefits of Biotech crops are well known and their cultivation expands, consumers' understanding of Biotech crops is improv-

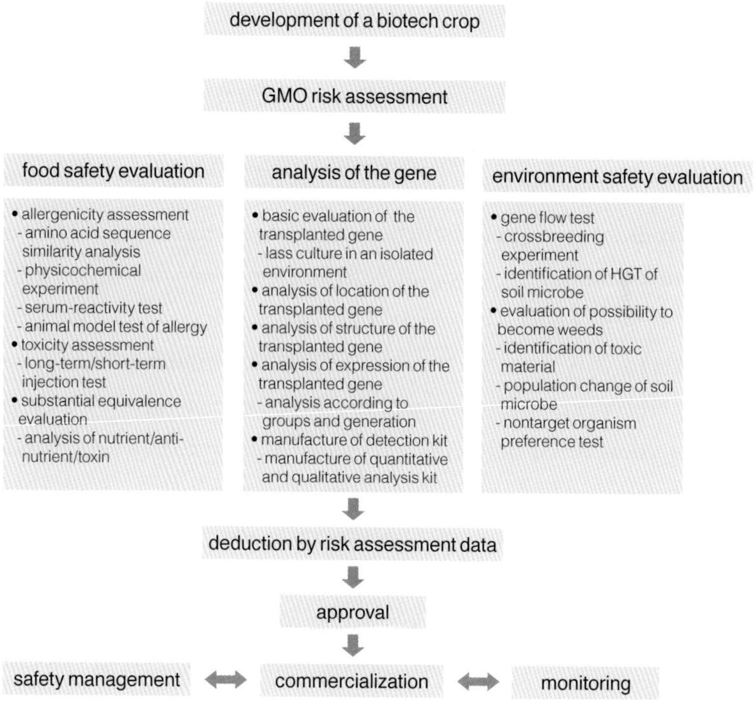

Figure 5-4. System and evaluation items of risk assessment for biotechnological crops

ing positively. Especially in Europe, where import of Biotech crops were banned, GMO products were actively accommodated since 2005; the cultivation of biotech crops is increasing and companies are developing new GMO products. Even though the beginning was slower than expected, the 21st century will be the era of Biotech crops. The potential of this technology will exceed our imagination if Biotech crops are used as bioreactors to effectively mass-produce expansive medicines and take place of fossil fuel to produce bio-energy and useful raw materials for plastics and other matters. At the end, it will become a survival technology for the absolute majority. Now it is time to work together to enjoy biotechnology safely and productively.

5-1. Risk Assessment of Biotech Foods

Food safety of newly developed Biotech crop is accurately assessed about novelty, allergenicity, antibiotic resistance, toxicity, etc. If certain substances of the food were something that were not present in the original crop or their contents were largely different, such food is considered new, and scope and degree for its test is suggested. It is important to check the foods -that are considered to have novelty- with toxicity tests for chronic toxicity and genetic toxicity and evaluate the results. However, there is no way to directly experiment on unexpected outcomes. Thus, scientists use data on original organism from which particular gene is extracted, the particular gene, transformation process, gene insertion site, and gene sequence of surrounding parts of inserted site. If protein produced by newly transformed gene is an enzyme, then there is a possibility of secondary metabolites being produced, and so it is reexamined. If the possibility is likely, toxicity tests on animals are done. Currently, fast-growing mice and chickens are used for testing, and any influence on them are collected and evaluated. However, if newly developed biotech product is similar to existing products in its components when compared, it can be considered as safe to human body by the concept of 'substancial equivalence'.

5-2. Food Safety Assessment Principle and Method

Food risk assessment is carried out based on universal standards such as precautionary principle, substantial equivalence, and familiarity principle.

Precautionary principle is, as its name indicates, discarding the product for preventative cause when its safety is questionable. Discard of Brazilian nuts well exemplifies such case. Current knowledge on food safety is empirically based on consumers' consumption rather than on scientific tests. When we analyze each component of food one by one, unhealthy substances related to hypertension, allergy, and dyscrinism are not small in number. So it is very difficult to define absolutely safe food. Consumers decide to choose less dangerous food or to bear the potential danger, based on their values and purposes. Such decision making process is based on accurate information, experience, and possibility of danger is largely considered.

Familiarity principle is cultivating biotech crops gradually just like cultivating ordinary crops with appropriate safety precaution and measures.

But in case of genetically modified food, there is no precedence of safe intake and empirical amount for safety. Even by the scientific assessment method, it is very difficult to accurately calculate the risk because no risk assessment for food itself has ever been conducted till these days. Method or tool to evaluate potential risk is limited and permissible range is also vague to determine. Therefore substantial equivalence means scientifically judge the safety by understanding

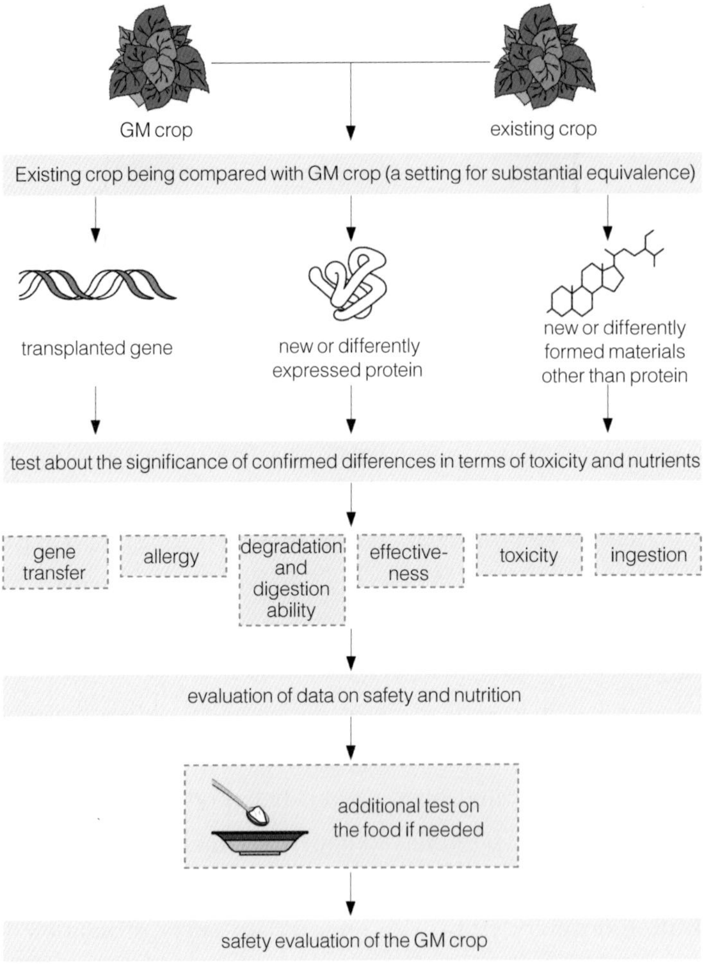

Figure 5-5. Flow of safety evaluation of biotechnological crops as foods

transformed gene's traits and comparing them with those of original food during the risk assessment of GM foods. Figure 5-5 shows the process of GM food risk assessment.

In order to assess the risk of food, scientists should know details about structures and traits of gene-donating organism, gene carrier,

Figure 5-6. Major allergy-inducing foods or factors

and of organism into which the gene transformed. Scientists examine whether toxins, nutrient inhibitory factors (anti-nutrients), and allergens are reported in gene-donating organism or accepting organism, then evaluate the possibility that safety problem or unexpected impact may emerge. Risk assessment test of desirable genes proves that a particular gene is safe when its gene sequence has no known harmful sequence. Scientists check whether transformed gene's structure and expression is stable through cultivation over generations, and evaluate the possibility that new problems may occur. They also need to check whether protein produced from transformed gene is similar in structure to known toxins, nutrition inhibitory factor, or allergens, and whether it is broken down fast enough by heat or artificial gastric juice and intestinal juice in cooking or processing step not to act as toxins or allergens.

Food allergic reaction is abnormal immune response to the sub-

Table 5-1. Ten major food that induce allergy

Researcher	Food
Kyu-Un Kim	Egg, pork, peach, mackerel, chicken, milk, buckwheat, crab, wheat, tomato
Eyermann	Wheat, egg, milk, beans, potato, fish, onion, tomato, cucumber, beer
Freinberg	Egg, wheat, milk, beans, nuts, fish, onion, potato, chocolate, fruits
Randolph	Corn, wheat, milk, egg, potato, orange, beans, beef, tomato, coffee
Solari	Milk, wheat, egg, rice, tomato, chocolate, beef, strawberry, oranges, chicken
Speer	Milk, chocolate, corn, oranges, egg, beans, tomato, wheat, apple, coffee

stance that is not problematic to most people. Food allergy is usually expressed to particular person in particular period of his life, but the cause is unknown. The cause for permanent allergic reaction to nuts and sea food is also not known. Diary, eggs, soybeans, grains, etc. are main cause of allergy for children whereas nuts, peanuts, clams, etc. are main causes for allergy in adults; reason for such phenomenon is not known, too. Generally 0.3~0.7% of population is known to suffer allergic reaction and the frequency is higher in younger age. Actually, around 70 children die every year because of peanut allergy. This report shows how common the food allergy is.

For these reasons, it is necessary to predict and evaluate the possible allergic reaction before we make food from GM crops. Bioscientists compare the protein sequence of gene which they are going to use with the allergenic protein reported. If more than eight amino sequences are identical or 35% of 80 amino sequences are homologous to the reported allergens, it is considered to cause allergic reaction and stop development.

As shown in previous example of Brazilian nut, the evaluation process is widely used; When Brazilian nut's gene was transplanted in order to increase nutrient content in soybeans, allergen in Brazilian nuts was also found in genetically modified soybeans in development

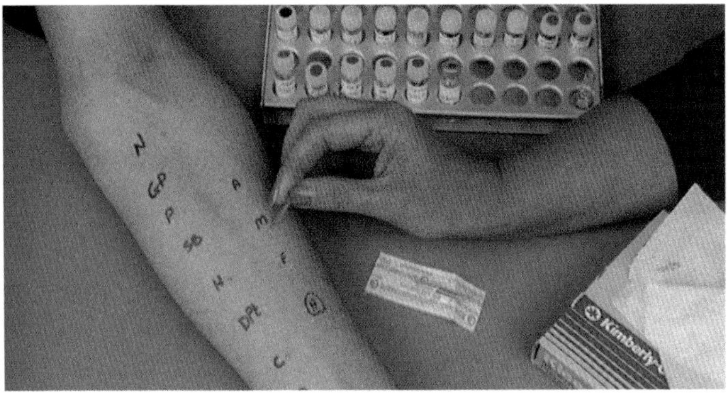

Figure 5-7. Allergic reaction test

step, and the whole development and research was shut down.

Antibiotics resistance problem is raised because we use antibiotic-resistant gene as selection marker gene in developing process of biotech crops. Possibility that antibiotics resistant gene of food may get transferred to microorganisms in intestine so the microorganism becomes resistant to antibiotics was raised in early stage of discussion on risk assessment for biotech crops. But such possibility is shown to be very rare.

In case of soybeans, there are about 200 mg of gene, the DNA, in every 100 g, and transformed gene is about 1/10,000,000 of the total DNA. Antibiotic-resistant gene or protein made from such a gene can function as just nutrients because they break down due to the digestive enzyme or acidic gastric juice.

Furthermore, there is no clear evidence that gene of food get transferred to microorganisms. Even though gene actually gets transferred to microorganism, it is hard to expect that gene can survive. Micro-organisms have self-defensive mechanism that they break down and destroy the gene from outside source. We Koreans have been eating rice for more than 10,000 years, but still there are not any reports that gene of rice is found in human's cells or colon bacillus yet.

In the beginning of risk assessment, there had been research to see if there is any possibility that allergic reaction may increase because of influence from biotech crop with antibiotic-resistant gene on bacteria in intestine. But, as it is mentioned earlier, the possibility that gene taken in by eating to be transferred to microorganism is very unlikely, and there is no direct effect to increase allergic reaction.

Recently, scientists are working on biotech crops without marker or with safe marker instead of antibiotic-resistant gene in order to reduce the worries. If antibiotic-resistant gene is necessary, resistant gene that is not important clinically any more is used, and research and development of such a marker genes are actively carried out.

During the toxicity test, protein produced by transplanted gene is injected more than at least 1,000 times of daily intake to observe the effect. In case of pest-resistant corn, 3,283 mg of pest-resistant protein is injected per 1kg of mice for 14 days. It is equal to injecting 6.5 ton of pest-resistant corn at once to 100 kg animal (Average weight of one pig is almost 90 kg.). Nutritional evaluation is also done by injecting it to young chickens for 42 days. During this period, the chicken becomes almost 6~7 times heavier than its original weight.

For the possibility that unintended effect may occur due to the gene transformation, scientists must analyze each components, check whether there is any significant difference in its composition, and present reasonable explanation to argue that it is safe. Possibility that protein made by transplanted gene may influence plant metabolism and create harmful substances is determined by the protein's function, while additional test is also required by clearly analyzing the gene sequence of region around transplanted site.

Scientists comprehensively evaluate in many aspects: they establish characteristics of recombinant genes that can cause harmful effects,

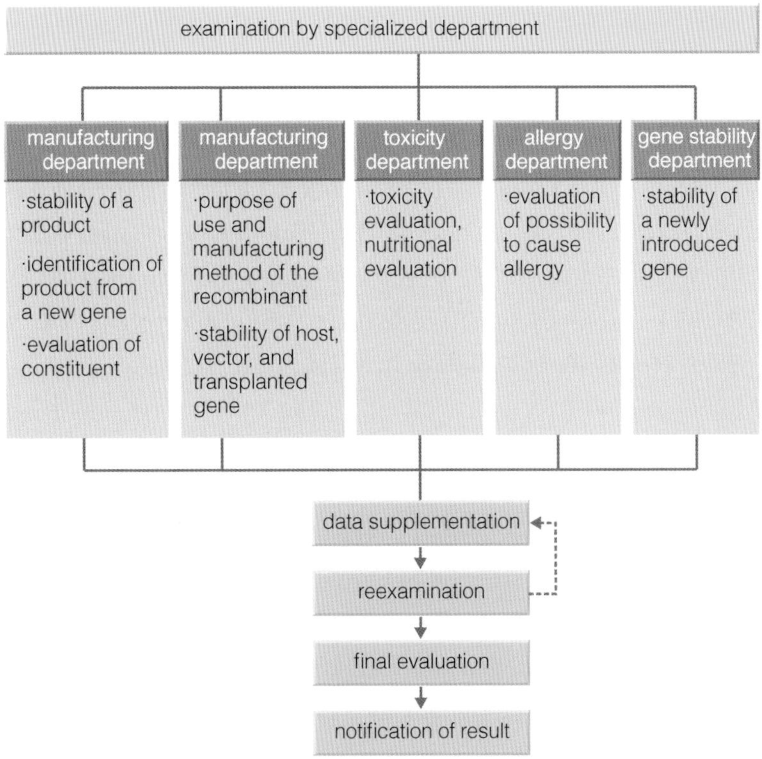

Figure 5-8. Flow chart of food safety examination

assess potential threat of each gene, and possibility how likely the threat is to actually occur.

5-3. Example of Food Safety Test of Biotech Crop in Korea

First product that passed the food safety test in Korea was Roundup Ready Soybean® made in 2002 which was able to grow without the influence of Glyphosate. Glyphsate, whose proprietary name is Roundup® and chemical name is phosphonomethylglycine, was first invented by American company Monsanto, and in Korean it is produced and sold by 12 different pesticide companies.

Until December 2008, total 7 crops 54 events of biotech crop product were approved of food safety test by Korea Food & Drug Administration: 1 event of Roundup Ready Soybean, 28 events of pest-resistant corn including MON810 (Monsanto), 13 events of cotton including herbicide-resistant cotton MON14445/1698 (Monsanto), 6 events of Canola including herbicide-resistant Canola MS1/RF1 (Bayer Cropscience) , 4 events of potato including RBMT21-129 which is resistant to Colorado potato beetle and potato virus Y, 1 event of sugarbeet including herbicide-resistant sugarbeet H7-1 (Monsanto), and 1 event of alfalfa including herbicide-resistant J101/J163 (Monsanto). Table 5-2 shows the details. When these are classified, there are 17 herbicide-resistant events, 10 pest-resistant events, 25 events with resistance to both herbicide and pests, and 2 pest-and-virus-resistant events. In case of corn and cotton, stacked varieties created by crossing genetically modified crops of different traits will be more common than genetically modified crops of a single trait in next generation, and the technology is generally spread.

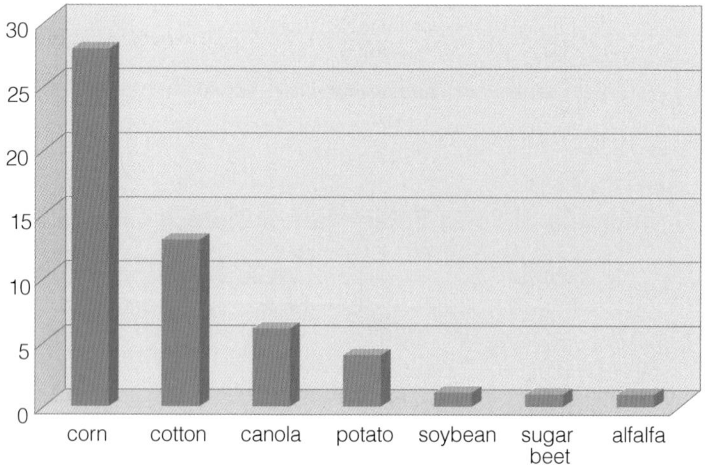

Figure 5-9. Status quo of registered crops in South Korea

Table 5-2. List of crops approved as edible goods by South Korean Ministry of Food and Drug Safety (December 2008)

No	type of crop	product name	property	insert gene	applicant
1	soy bean	PRS40-3-2	herbicide-tolerance (Glyphosate)	*cp4 epsps*	Monsanto
2	corn	MON810	insect-resistance (Corn Borer)	*crylAb*	Monsanto
3	corn	TC1507	herbicide-tolerance (Glufosinate) & insect-resistance (ECB, Hubner, SWCB, BCW)	*cry1F, pat*	Dupont
4	corn	GA21	herbicide-tolerance (Glyphosate)	*m-EPSPS*	Monsanto
5	corn	NK603	herbicide-tolerance (Glyphosate)	*cp4 epsps*	Monsanto
6	corn	Bt11	herbicide-tolerance (Glufosinate) & insect-resistance (Corn Root-worm)	*cry1Ab, pat*	Monsanto
7	corn	T25	herbicide-tolerance (Glufosinate)	*pat*	Aventis Crop-Science
8	corn	MON863	insect-resistance (Corn Rootworm)	*cry3Bb1, npt II*	Monsanto
9	corn	Bt176	insect-resistance (Corn Borer)	*cry1Ab, bar*	Syngenta
10	corn	DLL25	herbicide-tolerance (Glu-fosinate)	*cry1Ab*	Monsanto
11	corn	DBT418	herbicide-tolerance (Glufosinate) & insect-resistance (Corn Borer)	*bar, cry1Ac*	Monsanto
12	corn	MON863× NK603	gene stacked event (insect-resistance (Cotton Bollworm) & herbicide-tolerance (Glyphosate))	*cry3Bb1/ cp4epsps*	Monsanto
13	corn	MON863× MON810	gene stacked event (insect-resistance (Cotton Rootworm/ Corn Borer))	*cry3Bb1, cry1Ab*	Monsanto

No	type of crop	product name	property	insert gene	applicant
14	corn	MON810×GA21	gene stacked event (insect-resistance (Cotton Bollworm) & herbicide-tolerance (Glyphosate))	*cry1Ab/ mepsps*	Monsanto
15	corn	MON810× NK603	gene stacked event (insect-resistance (Corn Borer) & herbicide-tolerance (Glyphosate))	*cry1Ab/ cp4epsps*	Monsanto
16	corn	1507×NK603	gene stacked event (insect-resistance (ECB, Hubner, SWCB, BCW) & herbicide-tolerance (Glyphosate/ Glufosinate))	*cry1F/pat, cp4epsps*	Monsanto
17	corn	MON810× MON863× NK603	gene stacked event (insect-resistance (Cotton Bollworm/ Corn Borer) & herbicide-tolerance (Glyphosate))	*cry1Ab, cry3Bb1/ cp4epsps*	Monsanto
18	corn	DAS-59122-7	herbicide-tolerance (Glufosinate)	*cry34Ab1 and cry-35Ab1, pat*	Dupont
19	corn	MON88017	herbicide-tolerance (Glufosinate) & insect-resistance (Corn Rootworm)	*cry3Bb1/ cp4epsps*	Monsanto
20	corn	Bt10	insect-resistance (Corn borer) & herbicide-tolerance (Glufosinate)	*cry1Ab, pat*	Syngenta
21	corn	MIR604	insect-resistance (Corn Rootworm)	*mcry3A, pmi*	Syngenta
22	corn	Das-59122-7× 1507×NK603	gene stacked event (insect- resistance (ECB, Hubner, SWCB, BCW) & herbicide-tolerance (Glyphosate/ Glufosinate))	*cry34Ab1, cry35Ab1, cry1F/pat, cp4epsps*	Dupont

No	type of crop	product name	property	insert gene	applicant
23	corn	1507×Das-59122-7	gene stacked event (insect-resistance (ECB, Hubner, SWCB, BCW) & herbicide-tolerance (Glyphosate/Glufosinate))	*cry1F, cry34Ab1, cry35Ab1/ pat*	Dupont
24	corn	Das-591227× NK603	herbicide-tolerance (Glufosinate) & insect-resistance (Corn Root-worm)	*cry34Ab1, cry35Ab1, cry1F/pat, cp4epsps*	Dupont
25	corn	Bt11×GA21	gene stacked event (insect-resistance (Corn Rootworm) & herbicide-tolerance (Glyphosate/Glufosinate))	*cry1Ab/pat, mepsps*	Syngenta
26	corn	MON88017× MON810	gene stacked event (insect-resistance (Corn Rootworm/Corn Borer) & herbicide-tolerance (Glyphosate))	*cry3Bb1, cry1Ab/ cp4epsps*	Monsanto
27	corn	Bt11×MIR604	gene stacked event (insect-resistance (Corn Rootworm) & herbicide-tolerance (Glufosinate))	*cry1Ab, mcry3A/pat*	Syngenta
28	corn	Bt11×MIR604 ×GA21	gene stacked event (insect-resistance (Corn Rootworm) & herbicide-tolerance (Glyphosate/Glufosinate))	*cry1Ab, mcry3A, mepsps*	Syngenta
29	corn	MIR604×GA21	gene stacked event (insect-resistance (Corn Rootworm) & herbicide-tolerance (Glyphosate))	*mcry3A/ mepsps*	Syngenta
30	cotton	531	insect-resistance (Cotton Bollworm)	*cry1Ac, npt II*	Monsanto
31	cotton	757	insect-resistance (Cotton Bollworm)	*cry1Ac, npt II*	Monsanto
32	cotton	1445	herbicide-tolerance (Glyphosate)	*cp4 epsps, npt II*	Monsanto

No	type of crop	product name	property	insert gene	applicant
33	cotton	15985	insect-resistance (Cotton Bollworm)	*cry1Ac/ cry2Ab, uidA*	Monsanto
34	cotton	281/3006	herbicide-tolerance (Glufosinate) & insect-resistance (Corn Rootworm)	*cry1F, cry1Ac, pat*	Dow AgroSci-ences
35	cotton	15985×1445	herbicide-tolerance (Glufosinate) & insect-resistance (Corn Rootworm)	*CP4 EPSPS*	Monsanto
36	cotton	531×1445	gene stacked event (insect-resistance (Cotton Bollworm) & herbicide-tolerance (Glyhposate))	*cry1ac/ cp4epsps*	Monsanto
37	cotton	LLcotton 25	herbicide-tolerance (Glufosinate)	*bar*	Bayer
38	cotton	MON88913	herbicide-tolerance (Glyphosate)	*cp4 epsps*	Monsanto
39	cotton	Bollgard II 15985× MON88913	gene stacked event (insect-resistance (Cotton Bollworm) & herbicide-tolerance (Glyphosate)	*cry2Ab/ cp4epsps*	Monsanto
40	cotton	BG2×LL	gene stacked event (insect-resistance (Cotton Bollworm) & herbicide-tolerance (Glufosinate))	*cry2Ab/bar*	Bayer
41	cotton	281/3006× 88913	gene stacked event (insect-resistance (Cotton Bollworm) & herbicide-tolerance (Glyphosate/Glufosinate))	*cry1Ac, cry1F/pat, cp4epsps*	Dow AgroSci-ences
42	cotton	281/3006×1445	gene stacked event (insect-resistance (Cotton Rootworm) & herbicide-tolerance (Glyphosate))	*cry1Ac, cry1F/pat, cp4epsps*	Dow AgroSci-ences
43	canola	GT73	herbicide-tolerance (Glyphosate)	*cp4 epsps, gox*	Monsanto
44	canola	MS8/RF3	herbicide-tolerance (Glufosinate)	*bar, barnase barstar*	Bayer

No	type of crop	product name	property	insert gene	applicant
45	canola	T45	herbicide-tolerance (Glufosinate)	*pat*	Bayer
46	canola	MS1/RF1	herbicide-tolerance (Glufosinate)	*bar, barnase barstar, neo*	Bayer
47	canola	MS1/RF2	herbicide-tolerance (Glufosinate)	*bar, barnase barstar, neo*	Bayer
48	canola	Topas 19/2	herbicide-tolerance (Glufosinate)	*pat, neo*	Bayer
49	potato	SPBT02-05	Resistance to Colorado potato beetle (*Leptinotarsa decemlineata*)	*cry3A, neo*	Monsanto
50	potato	RBBTO6	Resistance to Colorado potato beetle (*Leptinotarsa decemlineata*)	*cry3A, npt II*	Monsanto
51	potato	RBMT15-101, SEMT15-02, SEMT15-15	Resistance to Colorado potato beetle/Potato virus Y	*cry3A, npt II, PVYcp*	Monsanto
52	potato	RBMT21-129, RBMT21-350, RBMT22-82	Resistance to Colorado potato beetle/Potato virus Y	*cry3a, PLRV rep, npt II, cp4 epsps*	Monsanto
53	sugar beet	H7-1	herbicide-tolerance (Glyphosate)	*cp4 epsps*	Monsanto
54	alfalfa	J101/J163	herbicide-tolerance (Glufosinate)	*cp4 epsps*	Monsanto

a. Judgment on Substantial Equivalence

RRS soybean was determined whether it is substantially equivalent to conventional one by data about genetic material, extensive eating experience, components and composition of food, and different ways of intake. First, in case of genetic material, host is soybean (Glycine mas L. cv. A5430), donator DNA is herbicide resistant EPSPS gene originated from soil microbe Agrobacterium sp. CP4, and NPTII gene for selection is isolated from E. coli. For the extensive eating experience, soybean has long been used as food, and Agrobacterium sp. CP4 is also considered safe because even though Agrobacterium sp. CP4 has not been intentionally eaten but crops infected by the bacteria was safe to eat. Furthermore, CP4 EPSPS protein's expression level is 0.239(0.179~0.303) μg per 1 mg of edible tissues, and NPTII protein is already proven to be safe becasuse it is registered as GRAS (Genetically Recognized as Safe) in USA. In case of food composition, l soybean's essential nutrients (protein, amino acid, ash, fat, fiber, carbohydrate, fatty acid, calories, etc.) and nutrition inhibitory substances (trypsin inhibitor, lectins, phytosterol, stachyose, raffinose, phaytate, etc.) was not different from that of conventional soybean regardless of glyphosate application. Based on these facts, herbicide resistant soybean was considered to be equivalent to conventional soybean and thus approved to be safe compared to safety assessment data of conventional soybean.

b. Purpose of Development and Its Usage

In reviewing safety test of transformant the validity of its purpose of development and its usage is first checked. Herbicide resistant soybean can use glyphosate during cultivation since gene that expresses CP4 EPSPS protein, which is not affected by glyphosate, is transplanted. Conventional soybeans need to deal with several selective herbicides in order to kill weeds, but herbicide resistant soybeans can use only glyphosate types when it is necessary. It was developed to reduce the energy to remove weed and cultivation, make easier to check residual concentration of herbicide, and decrease the use of herbicide.

c. Details about Host

Glycine max L. is a leguminous plant which contains harmful bioactive substance such as trypsin inhibitor, lectines, phytosterol, stachyose, raffinose, phytate, etc, and causes allergic reaction along with milk and eggs. But it has been considered as safe till now and had been used in various ways as food.

d. Details about Vector

In RRS soybean development, plasmid PV-GMGT04 was used in which promoter including CaMV 35S enhancer, part of petunia CTP gene to express transit signal peptide that transports EPSPS protein to chloroplast, terminator of Agrobacterium tumefaciens' nopaline synthetase, and EPSPS gene of Agrobacterium sp. CP4 are recombined. Size of DNA is 10,511bp, and restriction map and gene sequence is provided. It does not self-proliferate because it does not include harmful base sequence and has no transmissibility.

e. Details about Transgene and Its Product

CP4-EPSPS gene, a gene introduced as transgene, originates from Agrobacterium sp. strain CP4 which is soil microbe that contains many plant pathogens but there is no report on any disease or toxicity to human body. Particle gun method is used to transduce genetic cassette into the host. Insertion site does not contain harmful DNA sequence reported in toxin and allergen database. Stability of transgene (including change in gene recombinant) is shown by F2 segregation ratio by the crossbreeding of 3:1 indicating that is controlled by a single dominant allele in the nucleus. DNA analysis proves that gene is stable over four generations. Therefore, transgene is well inherited monogenically to next generations according to Mendel's law. According to genetic segregation ratio, recombinant gene is maintained in one locus. According to results on expression site, expression time, and expression level of recombinant gene were usually 0.046~0.798 μg/mg in leaves and 0.179~0.359 μg/mg in seeds by wet weight, even though some variations within and among regions were observed. Other than that, existence of another open reading fame and possibility of its transcription and translation was unlikely except for CP4-EPSPS gene considering the DNA sequence of trans-gene.

f. Details on Transformant

The new characteristic given by gene transformation is resistance to herbicide glyphosate through expression of EPSPS protein., No other toxic effect was detected in animal test. If the genetic product affects metabolism pathway, primary enzymes involved in metabolism are expected to change. But no changes seem to have happened based on the data. Also, the result of an extensive test on differences between host and transformant (data on nutrients, nutrition inhibitory factors, and harmful substances due to the changes in their contents) and nutrition and anti-quality substances of herbicide resistant soybean indicated that there was no significant difference between the host and the transformant.

Even though the period that food products containing gene CP4-EPSPS transplanted to RSS was approved in other countries and eating history are rather short, so far there has been no report that products of transplanted gene become allergens. Data on sensitivity to physicochemical processing of gene products states that CP4-EPSPS easily lose enzyme activity after heat treatment. It was easily degraded in digestive fluid in a test tube and very sensible to physicochemical treatment. Also, according to data on the structural similarity to well known food allergens and search results on structural homology to known allergens by 1992, no homology was reported.

Whether the amount of gene product occupies significant portion in daily protein intake is explained as follows. About 0.02 g of

CP4-EPSPS protein is expressed per 100 g of soybean According to '95 National Nutrition Survey Report of Korea ('95. 3) Korean intake average 3.6 g of soybeans everyday even though it varies about 3.2~4.6 g depending on regions. Also, the average intake of processed soybean such as tofu and soy milk is less than 29.3 g (5g as soybean) and thus total daily intake of soybean is less than 8.6 g. If we assume that they are not processed and there is no degration of protein, then the protein intake is presumed to be about 2 mg. Such a amount is about 0.0027% of 73.3 g of Korean's daily protein intake, and CP4-EPSPS is sensitive to physicochemical changes and is easily broken down and changes. When the assumed daily intake is compared with other food allergens, it is much less than 260 mg of milk's ß-lactoglobulin (about 0.58%) and physicochemically sensitive. Therefore it is hard to think that CP4-EPSPS causes allergic reaction. Also, when the serum of a person who is allergic to soybean and

Table 5-3. Food Safety Evaluation Items

1. Outline of evaluation requested food
2. Examination of suitability as food
3. Safety of genetic recombinant
    - A. Purpose of genetic recombinant development and data on the way of utilizing it
    - B. Data on the host
        - (1) Taxonomic characteristics (general name, scientific name, phylogenesis, etc.)
        - (2) History of cultivation and improvement by breeding
        - (3) Known toxicity or allergenicity
        - (4) History of safe intake
    - C. Data on the donor
        - (1) Taxonomic characteristics (general name, scientific name, phylogenesis, etc.)
        - (2) History of safe intake

(3) Donor and related species' toxicity, anti-nutrients, allergenicity relation to pathogenicity and known pathogens in case of microorganism)

D. Data on genetic recombination

(1) Data on trait change

(a) Trait change method (Agrobacterium method, particle gun method, protoplast method, etc.)

(b) Data on vectors used in recombination

1) Origin

2) Identification in host

3) Function in host

(c) Data on intermediate host

(d) Data on transmissibility

(2) Data on Introduced gene

(a) Characteristics of recombinant genes

1) Selection marker

2) Regulatory elements

3) Other factors affecting the function of DNA

(b) Size and Name

(c) Location and direction of gene sequence in complete vector

(d) Absense of harmful base sequence

(e) Absense of another open reading frame and its possibility of transcription and expression

(f) Absence of unnecessary sequences other than target gene (Purity of gene)

E. Data on characteristics of genetic recombinants

(1) Data on introduced gene in genetic recombinant

(a) Characteristics and function of the gene inserted into the genome of genetic recombinant

(b) Number of insertion site

(c) Structure of inserted gene at each inserted site

1) Copy number, base sequence (including flanking base sequence)

2) Data proving that there is no such genes encoding known toxin or anti-nutrients

(d) Absence of another open reading frame in inserted genes and flanking region of host genome, and its possibility of transcription and expression

(e) About safety

1) Sequence and size of inserted gene after multi-generation

2) Expression site, expression time, and expressed level after multi-generation

(2) Data on gene product

(a) Chemical characteristics of gene product (protein or non-transcribed RNA)

(b) Function of gene products

(c) Post-translational modification of expressed protein after its translation

(d) Structural changes of expressed protein

(e) Phenotype of new characteristics

(f) Expression site and expression level of gene products

(3) Toxicity

(a) When the product is a protein

1) Safe intake experience

2) Similarity of amino acid sequences to known toxins and anti-nutrients

3) Sensitivity of gene product to chemico-physical treatments

4) In case of the protein without any safe intake experience, oral toxicity test and biological function of the protein in the organism from which it was derived

(4) Allergenicity

(a) Data on whether gene product is known as allergen

(b) Sensitivity of gene product to chemico-physical treatments (in case of alternate product, include data on biochemical, structural, functional similarity)

(c) Data on homology to known allergens of gene product

(d) Data on whether the geneproduct account for significant amount of acceptable daily protein intake

(e) Following data if it is difficult to determine allergenicity based on (a) or (d)

(5) Difference from host

(a) Major nutrients

(b) Micronutrients

(c) Endogenous toxins

(d) Nutrition suppressor (anti-nutrients)

(e) Metabolites of inserted gene

(f) Nutrition

(6) Effect of gene product on metabolic pathway (possibility of reaction ong inherent substance of host as substrate)

(7) Current state of food distribution approval and intake in foreign countries

the serum of a person who is allergic to peanuts were analyzed and compared immunologically, no difference was found. From these results, additional allergic substances of herbicide resistant soybeans was determined not to be problematic.

As data on constituents, major nutrient content of fresh seeds such as protein, fat, fiber, ash, and carbohydrate were compared with those of conventional seeds for 2 years, the result was not different. In protein composite or fatty acid content, there was not big difference in major fatty acid content like palmitic acid, stearic acid, oleic acid, and linolenic acid. Just C22:0 fatty acid, which is contained very small amount in soybean (<0.6%), has significant statistical difference, but it does not exceed the deviation range generally stated in books that there is no biologically significant difference. On the other hand, between herbicide resistant seeds and fresh seeds there is no regardful difference in content of intrinsic toxin such as lectin, iso-flavin, stachyose/raffinose, phytic acid, and trypsin inhibitor.

EPSPS protein, a metabolic product of transplanted gene, is not identical or similar to any other toxins or allergens in sequence even though existing database has been searched. No toxic reaction was observed even though 562 mg/kg, which is 1,300 times of normal EPSPS protein expression amount, of CP4-EPSPS protein produced by E.coli is injected into rat's mouth. As the protein was broken down very quickly in artificial digestive fluid (within 15 seconds in gastric juice), the possibility of being absorbed at mucus of small intestine is very low. As expression amount of CP4-EPSPS protein in herbicide resistant bean was about 0.02~0.03% of weight of soybean and is ≤0.1% of total weight of soybean, it cannot affect the total amount of daily protein intake and thus doesn't seem to be problematic.

In the animal experiment, food intake rate and weight increase rate were given as a nutritional test data. Herbicide resistant seeds and conventional seeds were given as a feed to mice, cows, and poultry for 4 weeks before or after heat treatment. There was no big difference in animal's growth rate, and wholesomeness as a feed based on animal feeding test (using white rat, chicken, milk cow, catfish, and quail.)

5-4. Evaluation of The Biotech Crops on Environmental Safety

When cultivating biotech crops for a long time, it is crucial to accurately evaluate the effects of these plants on the environment and prevent them from disturbing the ecosystem. Also, it is necessary to make sure whether there is any possibility that the transformed gene will move to another different species and contaminate it, and the possibility of mutant's emergence. Based on the accurate and objective criteria on whether the biodiversity is affected, environmental safety of biotech crops is evaluated thoroughly.

The environmental benefits of biotech crops are as follows; 1) the productivity of crops can be enhanced by reducing the amount of chemical pesticide, 2) the undeveloped farmland that has outstanding quality of ecology can be preserved as natural ecosystem because of the increased production of crops leading to decline in the necessity of developing new farmland, 3) pollution in the Earth can be reduced through economical production of eco-friendly alternative energy such as bio-ethanol and bio-diesel through biotechnology.

On the other hand, a concern regarding the negative influence of biotech crops was raised. Because artificially inserted herbicide-tolerant genes might be spread to the weeds of other species through pollen and those that are modified as super weeds might not be removed by general pesticides from the ecosystem, the whole ecosys-

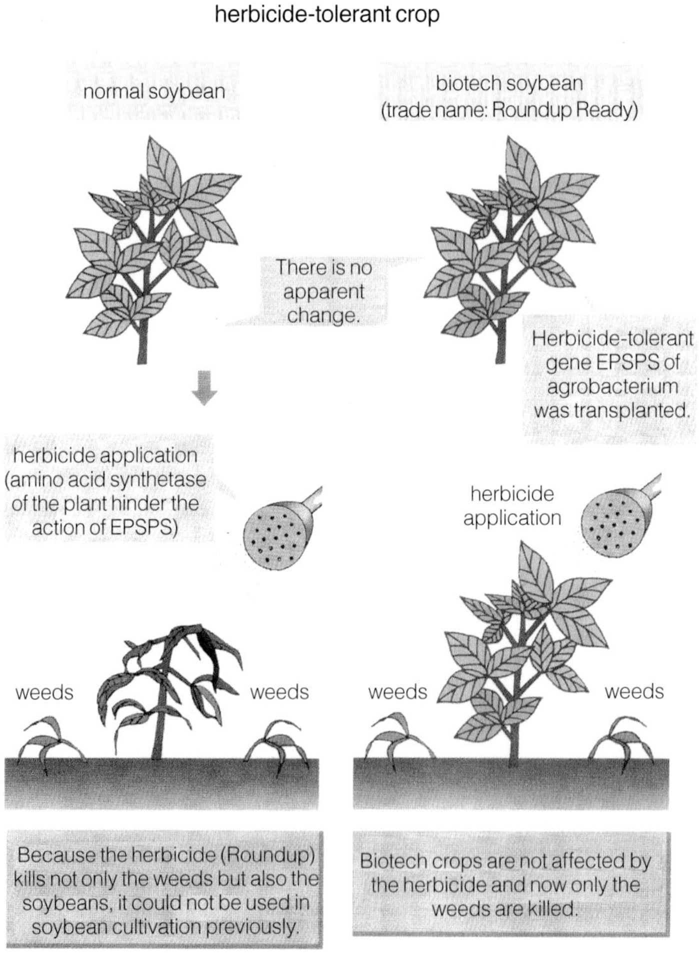

Figure 5-10. Effects of herbicide-tolerant crops

tem could be covered with those super weeds. Furthermore, there was another opinion that since biotech crops might have higher possibilities of viability compared to the conventional crops, only the biotech crops would survive in the future.

In order to test the possibility, British scientists carried out experiments about four biotech crops for 10 years and announced the results on the Nature, the most prestigious scholarly journal, in 2001. They researched the emergence, hybernation ability, and viability of the new potato, canola, corn, and sugar cane that contain transformed herbicides and insect insect-resistant genes in twelve areas. As a result, they were able to find out the difference between the biotech crops and the conventional crops. They reported that those transformed crops were rather behind in the competition with surrounding wild plants and were completely vanished in 4 years.

The insect-resistant crops are transformed with genes separated from biopesticide microorganisms that can selectively kill only larvae of lepidopteran insects, thus protect themselves from the pests. There was a worry that biotech crop that is resistant to particular pests can harm innocuous, beneficial, or even nontarget organisms, leading to the destruction of biodiversity. This started from a controversy that as the growing areas of insect-resistant biotech corns increased, larvae of the monarch butterfly could be harmed unexpectedly. After investigating the first experiment thoroughly to resolve this controversy, it was found out that the larvae died not because of the biotech crops but because of the inappropriate experimental environment.

Actually, in 2001, six research teams of United States and Canada implemented a large scale detailed experiment and found out that the larvae of monarch butterfly cannot be harmed by the pollen of pest resistant corns and the result was reported to the Proceeding of National Academy of Science, USA. It was shown that the expression of pest resistant gene in biotech corn pollen is so weak that the larvae of monarch butterfly hardly have a chance to be exposed to danger. They concluded that there are almost none of the unintentional harmful effect regarding the distance from the corn field, the period when pollen is blown away, and the period when the larvae

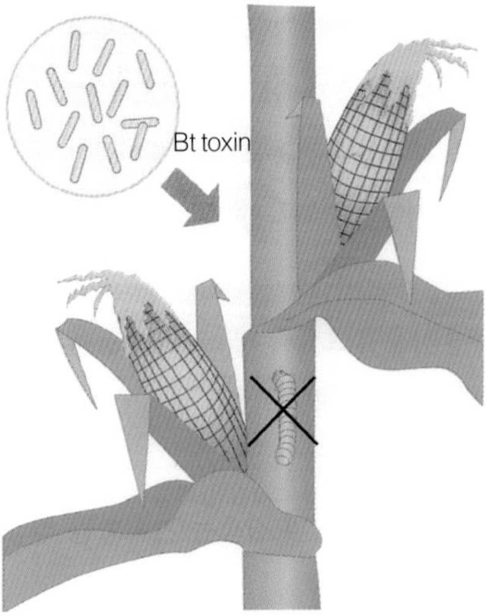

Figure 5-11. Insect-resistant corn

are growing. US Environmental Protection Agency, based on those experiment results, announced officially that the growth and development of the monarch butterfly is not affected by the cultivation of pest resistant corns.

On the other hand, when biotech crops are cultivated, pests such as corn borer, Ostrinia furnacalis, might acruire resistance and as generations go on, density of those pests inside the group of insects would increase. It might leads to a high possibility of all *Ostrinia furnacalis* in the ecosystem acquire such resistant traits towards insect-resistant genes. Thus EPA recommended in 1996 to plant conventional crops in certain areas as refuges when cultivating pest resistant biotech crops. Through this, the emergence rate of pests with resistant genes would be decreased. In the case of biotech crops that were transformed with two different pest resistant genes simultaneously, pests were known to have almost no possibilities to acquire such resistance.

There was another worry that the increased usage of herbicide resistant crops can rather increase the overuse of herbicides. However, in the past 10 years, the amount of herbicides actually decreased more than 14% in different regions. As cultivated areas expand, decline in the amount of herbicides is seen to be accelerating in the future.

The life scientists have tried hard to decrease the risks of biotech crops through such various researches and methods and have also tried to find the risks associated with the biotech crops. Specific risks have not been proven in the field yet.

Controversy regarding the safety of biotech crops is a natural process trying to find the point of social consensus while new science or technology is being industrialized. So, it is natural to go through thorough verification. If the benefits brought by the biotechnology towards humans and the environment are bigger than the risks, it can be naturally industrialized.

The risk assessment of biotech crops on the environment has complex system since it is very variable according to the traits of genes that are introduced and also according to the host organisms that the genes are introduced and the environment that they are released into. Thus, different from conventional evaluation systems of chemicals on environmental risks, biotech crop requires additional items for case-by-case system according to different traits of biotech crops.

5-5. The Rules and Methods of Environmental Risk Evaluation

In case of biotech crops, evaluations of environmental risks are carried out for the purpose of evaluating the negative effects that modified crops might bring into the ecosystem and preventing the outflow of these crops as food or feedstuffs until the environmental risks are assessed. So the developers of the biotech crops have to get environmental risks assessed through each country's agencies before the commercialization and mostly get assessed on the potential of environmental risks through the process in closed environment experiments.

The major content of the qualification of biotech crops is to evaluate the potential environmental influence of genetically modified organisms as mentioned previously such as disturbance of evolutionary process, destruction of biological diversity, biotech crops becoming weeds or emergence of new resistant diseases and pests, gene transfer to similar or non-target organisms, and negative influence towards secondary or non-target organisms.

Biotech crops are living organisms, different from toxic chemicals, which are not diluted in the ecosystem, and can form new living environment, proliferate, and produce metabolites continuously. Therefore when develope biotech crops, not only human body risks but also environmental risks should be cared for, and it is necessarily required to adopt the technology that can minimize the potential risks, scientifically and reasonably evaluate the developed biotech crops, and monitor it afterwards.

Environmental risks evaluation for agricultural biotech crops are supervised by the Korean Agricultural Biosafety Information Center of Rural Development Administration(http://kabic.naas.go.kr).

Table 5-4. The Risk assessment items for biotech corns

1) Theoretical basis for the development
2) Utility and the usage of development
3) Host
  a. Taxonomy (scientific name, general name, cultivars and lines)
  b. Distribution in the nature
  c. Usage by human kind (including usage abroad)

d. Biological traits (including survival, reproduction, and genetics)

(1) crossability

(2) Outcrossability with wild species

(3) Reproducibility with cultivated soybeans

e. Possibility to produce harmful materials

f. Contamination by Patogens or other harmful agents

g. Reproduction method and genetic traits

h. Place of origin, genetic traits and and other important physiological properties

i. Parasitism, sedentariness and other important physiological properties

j. Possibility of weediness

4) Foreign DNA donor organisms

a. General name and taxonomic traits (including scientific name, cultivar, and line)

b. usage by humans

c. Biological characteristics

d. Possibility to produce harmful substances

5) Vector

a. Name and origin (GenBank accession NO, etc)

b. Properties

(1) DNA molecular weight

(2) Restriction enzyme maps

(3) Presence of harmful base sequence

(4) Copy number and stability in the host

(5) Functions and traits

c. Pathogenicity

d. Information about the composition of vector

e. Tolerance to antibiotics

f. Presence of other marker genes and its propertities kind

6) Inserted genes

a. Functions and characteristics of the inserted genes

b. Origin and base sequence of each compartment of inserted genes

c. Details of modifying genes for usage

7) Methods and traits of genetically modified plants

a. Method of gene modification

b. Explanation about the process of fostering genetic modified plants

c. Future stability of the dominant trait by the inserted genes

8) Agricultural traits of genetically modified crops

a. characteristics and properties improved after modification
b. difference from the host or species that the host belongs to
(1) Reproduction method and genetic traits
(2) Characteristic of weediness
(3) Possibility to produce harmful substances
(4) other important physiological, morphological, and agricultural traits
c. information about the target substances and target organisms

9) Molecular biological verification of genetically modified plants
a. structure of the inserted genes of genetically modified plants
b. location of the gene inserted and its flanking sequences
c. copy number of inserted genes
d. Stability, expression and inheritance of inserted genes, with generations
(1) Stabilization of inserted genes in the future generation
(2) Expression of inserted genes
e. detection methods for inserted genes and its expression

10) Results from simulated environment experiment

11) Risk assessment of genetically modified crops
a. information related to production of toxic substances
(1) toxin from RR soybeans and evaluation of anti-nutrients
(2) production of allergens
(3) changes in major nutrient components
b. information related to possibility to become weeds
c. information about influence that can impose on surrounding organisms and ecosystem
d. information related to environment that the genetically modified plants are introduced to
(1) distance from the genetically modified plants to the origin
(2) information about geology, climate and ecological traits of neighboring plants

12) Approval status abroad and usage information

13) others (plan for implementing monitoring and methods, inactivation method for genetically modified plants, ways of handling in emergency such as accidents, etc)

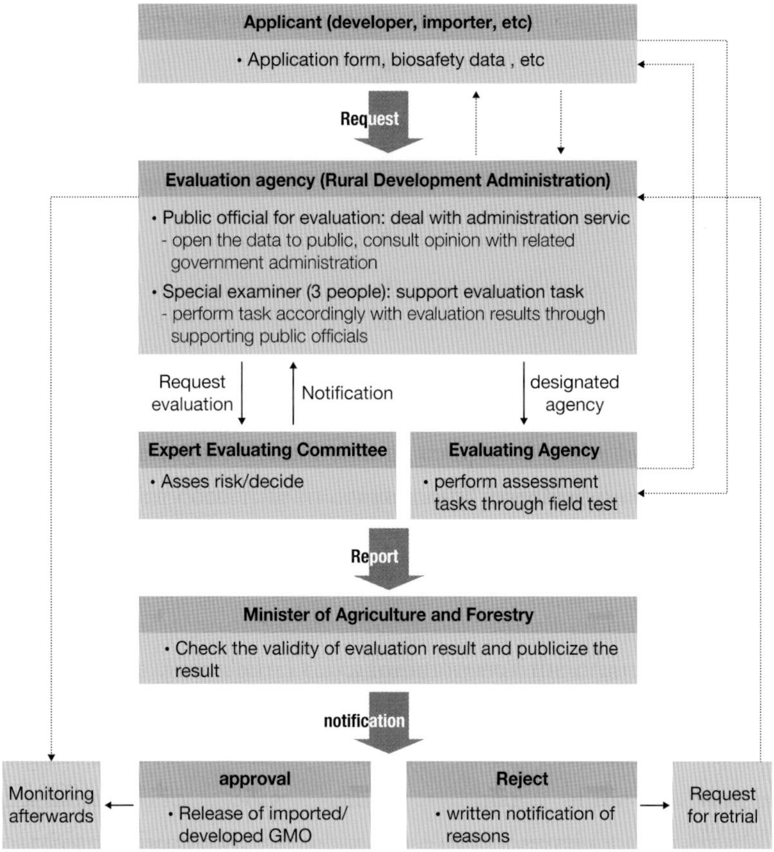

Figure 5-12. Environmental risk assessment procedure for genetically modified organisms

5.6 Approved Biotech Crops in Korea

Biotech crops in Korea that went through environmental safety evaluation and got approval for distribution from 2004 to 2008 reach 26 events for 5 crops (Table 5-5).

Table 5-5. Crops that were approved through genetically modified organism environmental safety evaluation (December, 2008)

Crops	Product name
Bean	*RRS 40-3-2* 1 item
Corn	*MON810, NK603, MON863, T25, TC1507, GA21, Bt176, Bt11, DAS-59122-7, MON88017, MIR604*, etc 11 items
Raw cotton	*RR1445, 531, 757, 15985, LLcotton25, MON88913, 281/3006*, etc 7 items
Canola	*T45, MS8-RF3, RT73, MS1/RF1, MS1/RF2, Topas19/2*, etc 6 items
Alfalfa	*J101, J163, J101XJ163* 1 item
Total	26 items for 5kinds

Data: Korean Agricultural Biosafety Information Center of Rural Development Administration (http://kabic.naas.go.kr)

Chapter Six

Safety Management Status of Biotech Crops

In order to secure the customers' health and environmental safety, the development and distribution of Biotech Crops undergo a rigorous control and management system, internationally and locally, with the developers' voluntary safety examination. It is considered as a necessary action until the historical safety is secured. The system has already been set by tracking down and analyzing different influences after the market distribution.

Safety management of genetically modified organisms is divided into several categories including: safety management of the environment, usage after the harvest, and safety management of humans or animals as related food or feedstuff. Internationally, the developed countries with OECD as a guide, discussed about the safety assessment rules according to commercialization, methods and managerial strategy before commercialization of biotech crops in the early and mid- 1990s and safety management system along with commercialization of biotech crops was established by the developed countries.

Discussion on environmental safety was handled mostly by OECD and UNEP, and in 2000 Cartagena Protocol on Biosafety was concluded based on the Convention on Biological Diversity (1992) mostly done by UNEP. As an implementation act related to this issue, South Korea also enacted "The act on trans-boundary movements of living modified organisms and other related matters (LMO act)" in 2001. In 2008, the Biological Safety Agreement came into effect, under the full-scale management system, and the Korea Ministry of Trade, Industry, and Energy arranged minor legislation to be recognized as a responsible agency in the national level.

Discussion on food safety is unrelated to UNEP, and mostly handled by CODEX, which is a jointly affiliated organization of WHO and FAO. While most of OECD member states have enforced safety management system of biotech foods in the late 90s, , the developing and underdeveloped countries are doing activities for the spread of management system with the CODEX Committee in the absence of their own independent system because it was difficult to establish their own regulation. One of the results from the CODEX Committee is the establishment of the 4 year Technical Expert Group from 2000 to 2007, which includes safety assessment rules and methods about biotech crops in 2002, biotech microorganisms in 2003, and food driven from biotech animals in 2007.

Other developed countries established environmental risk assessment systems according to the cultivation experiment and industrial cultivation in the early 90s and ultimately prepared food safety management system in the mid-90s. On the other hand, Korea built food safety assessment system in the early 2000s and then, implemented environmental assessment management system in 2008. In other words, Korea had implemented safety assessment as voluntary law under KFDA notification since 1999, and in 2004, amended the Food Sanitation Law, making food safety assessment mandatory. In order to accomplish competent environmental assessment, the Biological Safety Agreement came into effect, leading to the full-scale management system in 2008. Similarly, management system of biotech crops established as management technology is advanced accordingly with development speed of manufacturing technology, and furthermore, development of technology that solves the social issues on biotech crop development technology is done altogether. Under these systems, there is no case in which concerns came into real effect, even though the cultivation of commercialized biotech crops dramatically increased, and these developments of manufacturing technology and management technology will take place in the future.

6-1. South Korea

South Korea stated experimental guidelines and implementations of handling and using genetically modified organisms as found in the 'Biotechnology Support Act' enacted by the Ministry of Science and Technology in 1983 for the first time for basis establishment and activation of domestic biotechnology industry. After that, as safety problems of genetically modified organisms became an important issue internationally, the Ministry of Health and Welfare announced 'Guidelines regarding genetic recombination' in April, 1997. The main content included sealing standard, methods, storage, transportation, transfer, and disposal after the experiment of genetically modified organisms and mainly focused on pathogenic microorganism experiment.

In August of 1999, the Korea Food and Drug Administration (currently, Ministry of Food and Drug Safety) established 'Safety assessment guidelines for genetically modified food and food additives', and regulated that in the case where genetically modified organisms produced domestically are distributed as food, the developer or importer has to submit experimental data of nutritional contents, toxic substances, allergen, antibiotic resistance gene and has to be approved before distributing the product to the market. From February, 2004, it has been mandatory to go through safety assessment according to Food Sanitation Act. Following this guideline of KFDA, there was an assessment process of 'Roundup Ready Soybean', which is a herbicide-tolerant soybean, for approximately one year requested by Monsanto Korea and its safety was checked in 2002. After that, until December 2008, there have been 54 products for 7 different crops, including 'Roundup Ready Soybean' which went through the safety assessment process. The process is implemented continuously with imported crops. In the case of environmental safety assessment, it was started from late 2004 and there have been 26 products for 5 different crops that were approved until December of 2008.

In January, 2000, in Montreal, Canada, international standard 'Cartagena Protocol on Biosafety to the Convention on Biological Diversity' was adopted for the purpose of safety security of environment and humans in case genetically modified organisms are transferred between countries. As this occurred, Korea also established and announced the 'The act on trans-boundary movements of living modified organisms and other related matters (LMO act)' on the basis of the protocol by the Ministry of Commerce, Industry, and Energy in March, 2001. This law regulates safety assessment and examination procedures of imported or domestic genetically modified organisms, reporting of establishment and operation of experiment facilities for genetically modified organisms, items of license, and penalty clauses when violated. Also, it includes the enforcement decree and the enforcement rule based on LMO act. The protocol came to an effect internationally since September, 2003 when 50 countries deposited instrument of ratification and currently in June of 2008, 147 countries, including EU, has become member states, except major LMO production export countries such as United States, Canada, and Argentina. Korea deposited the instrument of ratification in October 3rd, 2007 and became the 143rd member state. The protocol was implemented in Korea since January 1st, 2008. Domestic Biosafety management system based on this LMO act is shown in the Figure 6-1.

The Ministry of Agriculture and Forestry controlling the import of base materials of agricultural product notified 'Environmental Risk Assessment Guidelines of Genetically Modified Agricultural Products' as of January 9th, 2002, the Ministry of Agriculture and

Forestry notification 2002-2 for securing the safety in the level of production and import of biotech crops. This guideline regulates; environmental risk assessment process of genetically modified organism and evaluating agency, management of evaluating agency, isolated packaging for domestic packaging experiment and managing method of genetically modified organisms, formation and management of expert assessment committee for the expertise, opening general information about assessment requesting modified organisms and gathering opinions. Furthermore, penalty provisions regarding the violation of genetically modified organisms' production and approval, management of research facilities and experiment rules are also established. Environmental risk assessment started since 2004 and there are 26 products for 5 different crops that went through the safety assessment until December, 2008.

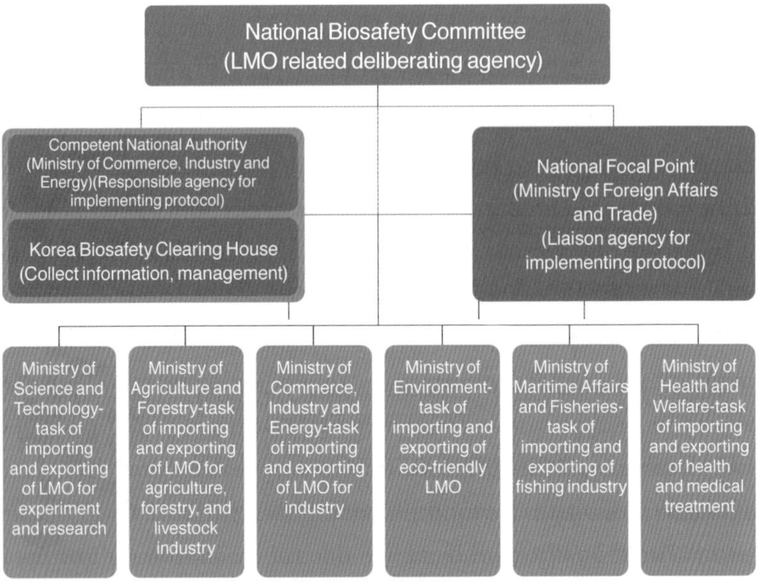

Figure 6-1. National biological safety management system based on LMO laws

The applicant, such as the importer, should make assessment application form and risk assessment data on a prescribed form and submit the documents to RDA (Rural Development Administration). Then, the expert assessment committee makes decisions after reviewing these documents. The assessment process is mainly focused on document evaluation and when needed, the agency can conduct packaging experiment and spot investigation. The evaluation process is to be finished within 270 days after applying, and when there is no risk, the result is announced to the applicant in writing. When the assessment result is rejected, demand for retrial is possible within 60 days after the decision date.

6-2. United States of America

The USA is a representative of a developed country with a long history of developing and testing the safety of GMOs. In 1986, the Office of Science and Technology Policy (OSTP) provided plans for resolving issues among agencies. These plans were proposed for the procurement of safety, effective management and supervision of biotechnological produce. Experts worldwide reviewed the safety issues of biotechnological produce, and concluded that it was possible to manage and supervise biotechnological produce according to the laws implemented in the status quo. Accordingly, these experts went on to review the 107 laws, rules, and guidelines already in existence. Additionally, debate took place on which laws were acceptable to apply on biotechnological produce, and which government branch would be responsible for what function. As a result, the following three organizations: Animal and Plant Health Inspection Service (APHIS) of the US Department of Agriculture (USDA), the Food and Drug Administration (FDA), and the Environmental Protection Agency (EPA) were decided to tend to these issues, and have been doing so for the past 15 years, gathering information and experience on safety management and supervision of biotech products.

The USDA is in charge of directing for imports and national movements of biotech crops and field trials and confirming that a new event of plant shows any difference from the original plant in a safety aspect. The objectives of regulations are plant diseases and pesticides, plants and veterinary products.

The FDA plays a role in monitoring the safety on food, feedstuff, food additives, and medical supplies and equipment for safety issues of humans and animals. The objectives of regulations are Micro-organism or pesticide, evaluating the environmental impact. The US Government puts top priority on food safety issues and has been proud of continually providing the world's safest food.

Before commercialization, biotech companies implement a very strict internal process on safety assessment and then, apply the safety assessment to USDA, FDA and/or EPA by the characteristics of GMOs. Approximately, new 130 GM crops have been developed and approved after the safety assessment. Thousands of food types, including GMO ingredients, are being distributed to the market.

The Function of the USDA and APHIS

The APHIS relies on the relevant laws such as the Coordinated Framework for Regulation of Biotechnology and the Plant Quarantine Act in managing and directing the possibility whether a new event can cause disease in the existing plants. However, a characteristic of the USDA is that its assessment begins in the very early stages of GMO research. During the initial stages of developing a new GM event, the USDA evaluates whether it can be safely used, and manages it to be used in agricultural performance trial or breeding program. On the other hand, safety assessments which are additionally requested for food and agro chemicals are decided upon through consultation with the FDA and EPA.

The USDA and APHIS mainly deal with the regulation on transportation of biotech crops between states, importation and the environment. Its tasks can be divided into two main categories: granting field trials and granting products. Safety assessments such as granting field trials are carried out by a Biotechnology Evaluation Team consisting of 19 experts, with each person being responsible for tasks such as release permits, notifications, and deciding petitions for non-regulated statuses.

A release permit is the most general permit, with the biotech crop being evaluated on environmental safety 120 days before being released into nature. If the assessment results turn out as FONSI (Finding of No Significant Impact), a release permit is granted. Unlike a release permit, a notification means that in the case of six types of produce (excluding certain products like weed) such as tomatoes, corn, tobacco, soybeans, cotton, and potatoes, environmental release trial can be allowed by just notification without being evaluated on environmental risk assessment. A deregulation petition is a method of granting which exempts the product from examination in order to accelerate the commercialization of the product. Once the product is deregulated, it and its progeny are exempt from regulations in importation, interstate transportation, and environmental release. They are treated as same as general seed.

The Function of the FDA

The FDA, based on the Federal Food, Drug, and Cosmetic Act, engages in determining whether a product is faulty or inauthentic. However, FDA keeps the position that specialized regulation in a biotech crop's production process is unnecessary. Its 1992 legislation proclaims that if a GMO does not create new toxins, modify existing toxins, or contain allergens, it is to be treated no differently from the original plant. Additionally, if there is a possibility that an already-approved GMO's nutritional contents can be transformed, allergens can be produced, or new cooking methods can be required, this information must be provided along with separate indication on the product. Such policies show, rather than new regulations or deregulation, the FDA's traditional stance on regulating GMOs.

The FDA's safety assessment lists clauses are as below.

(1) An unexpected creation or increase in toxins

Most plants self-produce toxins or anti-nutrients for self-defense purposes. However, crops, which have become refined under human hands, only produce toxins that are low in concentration, causing no harm to the human body. In the case of potatoes, there are no health issues because breeders have continually picked out plants with low levels of toxins. Even the potassium cyanide components in the ethanol material, cassava, are safe to eat after heated or processed.

Before commercialization, the developer must judge for certain whether or not the toxins in the GMO are within the scope of being harmless to humans. Certain proteins in produce may act as toxins, so the base sequence of the inserted gene must be compared with that of toxic proteins and checked for similarities. If similarities are found, the gene must be refrained from being used. If this gene is cultivated, it must undergo thorough safety assessments. Furthermore, to answer concerns of the gene being activated within the new events, mutations occurring as a result of evolution will soon stop the gene's effects. The problem is, an intermediate product may act as a toxin and be activated during the biotechnological process. However, there is a slim chance of this occurring with crops humans have been cultivating for centuries. Even if there were to be a possibility, it is likely that it would be discovered and removed during cultivation.

The GMO soybeans, corn, potatoes, and other products that have recently been confirmed for commercialization in the USA are crops that have been cultivated by humans for years, and have a history of being used as produce. Because of this they come in handy when determining the likeliness of a new breed generating toxins. If inspections on a crop's history of cultivation and consumption reveal a past case of toxin generation, the crop becomes the subject of rigorous food safety assessment before being commercialized. Even if the crop lacks a history of cultivation and consumption, it still undergoes the same assessment. This is because it is difficult to predict the likeliness of toxin generation in a new crop. The problem of toxins in food is not limited to toxins generated by the crop itself, but also includes fungous toxicity caused by secondary infection from pathogenic bacterium.

(2) Change in essential nutrients

Crops provide the human body with its most essential nutrients. For example, citrus fruits are a main source of Vitamin C, and carrots are the main provider of Vitamin A. According to the FDA's Food Safety Act, it is mandatory to compare a GMO's essential nutrient content with that of general seed, checking for differences. If the nutrient content shows to be significantly different, the related authorities may halt commercialization, or make it mandatory to add labels explaining the nutrient content changes in detail. In the case of GMOs that have been reinforced in iron or Vitamin A, a thorough food safety assessment must take place before commercialization. Labeling is mandatory even after assessment.

(3) Possibility of Causing Allergy

1~2% of the world's population suffers from alimentary allergy. These individuals show allergic reactions to one or two specific foods, and examples of such foods include: peanuts, soybeans, nuts, dairy products, eggs, fish, crustaceans, and wheat. Alimentary allergies are an issue because people with these may suffer sudden allergic reactions when consuming certain foods, and in the worst case, may even face death.

The FDA, in order to evaluate the allergy inducibility of GMOs, used the advice of experts in fields such as food safety, alimentary allergies, immunology, biotechnology, and diagnostics, and research results on alimentary allergies to create its safety assessment guidelines. According to these guidelines, in the case of biotechnologically developing one of the eight allergenic foods mentioned above, the developer must scientifically prove that there is no risk of generating allergic reactions.

All originally known allergens are proteins that contain a common amino acid sequence. Moreover, they share characteristics such as that their molecular weights are 10~70 kDa, or they are strong against acids or bases. Because of this, to ensure that the newly developed crop does not contain allergens, potential allergy tests must be held, determining whether the gene originates from an allergenic product, whether the gene-receiving plant has caused allergies in the past, and whether the introduced gene's protein sequence is identical to the originally known allergens. In one case, a pioneer hybrid seed company introduced Brazil nut genes to soybeans in order to strengthen methionine, an essential amino acid in soybeans. However, people reported allergic reactions against the Brazil nuts, and scratch tests and immunity tests showed a possibility of the reaction occurring in the biotech soybeans, as well. Consequently, development was discontinued.

(4) The Issue of Immunity to Antibiotics

Plant cells are highly inefficient for in gene insertion, and only a select few are used. Additionally, expression rates within the cells are extremely low. Mostly, an antibiotic-resistant gene is used as a marker to select the biotechnological cell out of millions. The cell containing recombinant genes is cultivated in an antibiotic culture medium, and a GMO is created through regeneration of the surviving celss. Concerns were raised on the possibility of an antibiotic-resistant gene being moved from the plant to pathogenic bacterium in the ecosystem, creating antibiotic-resistant pathogenic bacterium.

Accordingly, the FDA is evaluating safety according to the guidelines in its 1992 Manual of Policies, and its 1998 'Guidelines related to the usage of antibiotic-resistant genes in GMOs', shown below.

First, the possibility that the protein made from resistant genes is potentially toxicity

Second, the possibility that the created protein will cause an allergic reaction

Third, its importance as an antibiotic medicine

Fourth, whether or not the antibiotic is orally administered

Fifth, the characteristics of the antibiotic

Sixth, information on the possibility that the antibiotic-resistant gene will be moved from plant to microorganism, and whether or not the amount of antibiotic-resistant microorganisms has increased.

Such guidelines were decided upon by an expert group of microbiologists, medical scientists, bacteriologists, mycologists, and sitologists, and are used for food safety assessment not only domestically, but in international organizations and throughout. Although, there is no problem with the safety of antibiotic-resistant genes, current scientific technology is turning towards developing non-antibiotic markers, or not using markers at all. Such technology is already at commercialization stage.

The Role of the EPA

The EPA, based on the federal insecticide fungicide and rodenticide act (FIFRA), assumes the role of managing and directing the environmentally-safe usage of GMOs containing insecticide or herbicide-resistant substances. It also oversees the role of registration, labeling, and determining the conditions of using for new insecticidal substances. If an insecticidal substance is concluded to not cause any danger to public health, it may become completely exempt from the residue amount setup procedure. This is because as long as the biotech plant is not completely new, or has different aspects compared to the original plant, it has very low potential in causing negative effects to the human body and the environment. However, since 1999, when prominent academic journal *Nature* declared that study results indicated that Bt Corn pollen may kill Emperor Butterfly Larvae, a new restriction was established, making it mandatory that the developer hand in research guidelines along with research results that minimize effects on unspecific insects.

From 1987 to 2008, the total number of field trials was 13,900. Before 1900, the total was less than 65, but as full-fledged commercialization took place in 1995, the number radically increased, becoming approximately 1,000 per year by 1998, and maintaining a number of approximately 900 per year (Figure 6-2).

A total number of 57 species of GMO have had environmental release experimentation conducted more than 10 times. The top 9 crops – corn (6,380 cases, and 45% of the total amount), bean, cotton, persimmon, wheat, alfalfa, tobacco, canola, and rice in order

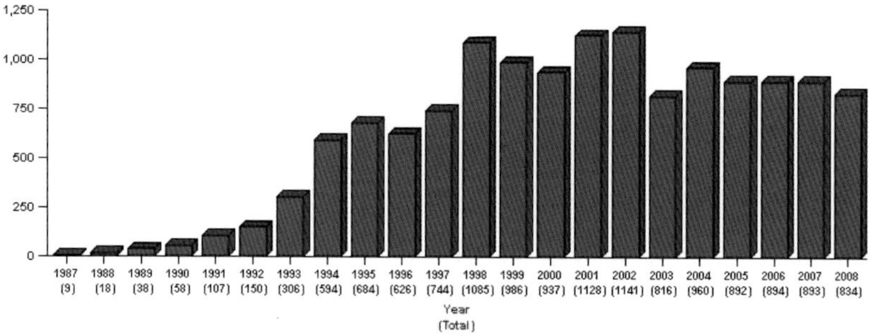

Figure 6-2. The number of granted environmental release field trials per year in the USA.

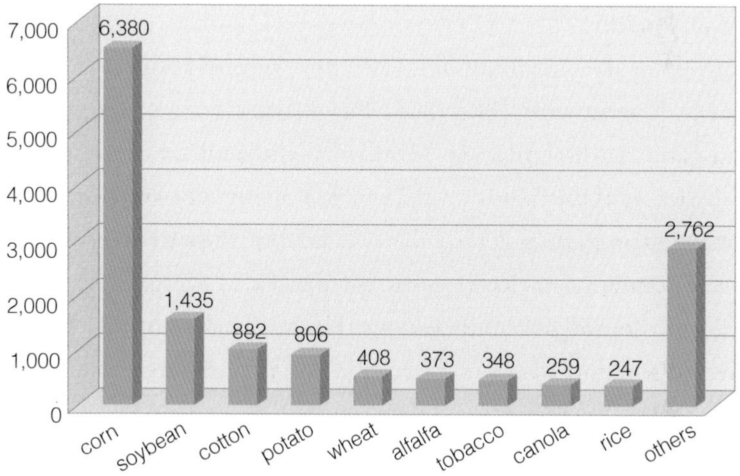

Figure 6-3. Status quo of granted field trials per crop in the USA

– take up 80% of the total amount. Of the characteristics under experimentation, herbicide resistance were the most, with 4,806 cases, followed by 3,933 cases on insect-resistance, 3,450 cases on enhanced agricultural product quality, 2,365 cases on enhanced agricultural characteristics, and 1,319 cases of virus resistance.

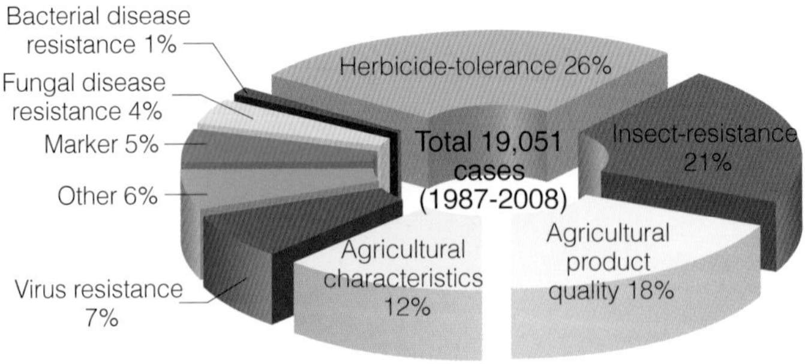

Figure 6-4. Characteristics distribution of crops granted field trial in the US

6-3. Japan

In Japan, four institutions - the Ministry of Education, Culture, Sports, Science and Technology, the Ministry of Economy, Trade and Industry (METI), the Ministry of Agriculture, Forestry and Fisheries, and the Ministry of Health, Labour and Welfare oversee the experimentation, commercial cultivation, importation, and safety assessment of food and feedstuff of GMOs.

Within Japan, one must submit relevant paperwork to three institutions- the Ministry of Education, Culture, Sports, Science and Technology, the Ministry of Agriculture, Forestry and Fisheries, and the Ministry of Health, Labour and Welfare- in order for both domestically developed and imported GMOs to be granted permission to conduct field trial, general field cultivation, and confirmation of safety as a food and feed. The amount of time before confirmation varies according to crop and characteristics, but generally takes approximately 5 years.

The Function of the Ministry of Education, Culture, Sports, Science and Technology

The Ministry of Education, Culture, Sports, Science and Technology accepted the OECD recommendations in 1987 and established its 'Guidelines for Recombinant DNA Experiments'. The guidelines include safety regulations in the case of conducting recombinant DNA experimentation. Moreover, it added the condition that gas emitted from the GMO be minimized, and the substances discharged from the working area be deactivated. It highlighted the responsibility of the individual, the continual training of experimenters, and the importance of the lab inspector. Licensees of a research institution must be fully aware of safety regulations and technology-related information, and are also responsible for the training of employees. Accordingly, it is mandatory that research institutions have a safety committee and a safety manager, and must devise plans to block infection when dealing with pathogenic microorganisms. The Ministry of Education, Culture, Sports, Science and Technology operates a safety committee consisting of 15 people, and it takes 1 to 2 months for an experiment related to recombinant gene technology to be approved. If it is decided that the applicant uses the facilities appropriately, an approval for experimentation is automatically granted, but until now there have been no cases in which experimentation was denied.

After gaining approval from the Ministry of Education, Culture, Sports, Science and Technology, it becomes possible to conduct closed and open greenhouse experiments. In closed greenhouses, the safety assessments are conducted on the expression of the introduced gene and morphological characteristics. In open greenhouses, safety assessments are conducted on morphological and growth characteristics, reproduction characteristics, and the possibility of toxin generation. When the assessments are over, isolated field trials become possible. The time until this happens is 2 years on average. Currently, at Tsukuba National Institute of Agrobiological Sciences, a BioPlant Research Center (BRPC) is operated for closed and open greenhouse experimentation. This facility is also used as a place where the general public may experience biotech crops.

In the case of domestically developed GMOs, it is mandatory to be evaluated by the Ministry of Education, Culture, Sports, Science and Technology, but in the case of imported GMOs, safety assessment experimentation is usually skipped, instead acknowledging the validity of safety experimentation results in the exporting nation.

The Function of the Ministry of Agriculture, Forestry and Fisheries

The Ministry of Agriculture, Forestry and Fisheries oversees issues related to the environmental release of GMOs. Crops that have been confirmed safe by the Ministry of Education, Culture, Sports, Science and Technology must then submit isolated field safety experimentation results to the Ministry of Agriculture, Forestry and Fisheries, according to the recombinant usage guidelines and food and feed additive safety assessment guidelines of agriculture, forestry, and fisheries. Assessment rubrics, information on experiment plan format, and isolated field usage plan format and relevant procedures are provided in order to evaluate the effect of biotech plants, microorganisms, and small lab animals in arbitrarily simulated environments or nature.

(1) Application for cultivation in simulated environment

The developer must fill in the greenhouse safety assessment results and a plan for the upcoming isolated field trial to apply for simulated environment cultivation experimentation. The recombinant usage special committee's 'recombinant plant subcommittee' primarily evaluates the plan, and evaluates the effect an isolated field trial would have on the surrounding environment. Applicants may be asked to provide supplementation upon need. The recombinant usage special committee reviews the assessment of the subcommittee, and reports to the Ministry of Agriculture, Forestry and Fisheries if there are no problems. The ministry secretariat's technology safety division has exclusive charge over granting simulated environment cultivation.

(2) Cultivation and Investigation in a simulated environment

In a simulated environment, it must be proven that the biotech crop has no significant difference in impact on the surrounding environment, and assessment points include wintering ability, shattering habit, and quiescency. Experiments occur in isolated fields and generally last one year. Seeds whose safety assessments have ended are incinerated, and strictly restricted from exiting the isolated field.

(3) Approval for general field cultivation and confirmation of importation

GMO developers or importers must attach the results for the environmental impact assessment experiment conducted in isolated field and send it to the ministry's technology safety division. There, they must apply as 'general field cultivation experimentation' or 'application for confirmation of importation'. When the paperwork is submitted, the recombinant usage special committee's 'recombinant plant subcommittee', consisting of approximately 15 experts in relevant fields, primarily evaluates the paperwork and may ask for additional supplementation on the experiment results. After assessment by the recombinant plant subcommittee is over, the paperwork is handed over to the recombinant usage special committee for secondary assessment, and then to the ministry's secretariat. When the secretariat's approval is confirmed, the biotech crop is now either able to be cultivated in general fields, or able to be imported. This process takes approximately 3 months.

The Function of the Ministry of Health, Labor, and Welfare

The Ministry of Health, Labor, and Welfare oversees issues related to medication and food. In 1987, it created a research group on the 'safety of foods and recombinant genes' that deals with safety issues related to the application of gene recombination. It especially conducts much research on the problems genetic engineering has created in society, and the general public's changes in opinion regarding it. Currently, the Ministry of Health, Labor, and Welfare has a consulation committee consisting of 12 people, overseeing the approval of GMOs as food products. Prior to 2001, no legal basis was held in safety assessments, but on April 1st, 2001, it became legally mandatory and the 'Food Additive Standard Regulations' were revised and proclaimed. The Ministry of International Trade and Industry operates a gene recombination technology committee comprised of ten members, and oversees the safety management of GMOs. It mainly deals with genetic engineering technology issues such as fundamental chemical substances, chemical products, and chemical fertilizer. It also deals with safety issues such as minimizing the fermenter's solution and release of the GMO. Safety regulation points include the gene recipient organism's classification, genetic characteristics, pathogenic or metabolic characteristics, the structure and method of the genetic recombinant molecule, the DNA of the gene and characteristics of the gene's original organism, characteristics of the gene expression in the GMO, and the similarity between the gene donor and the GMO.

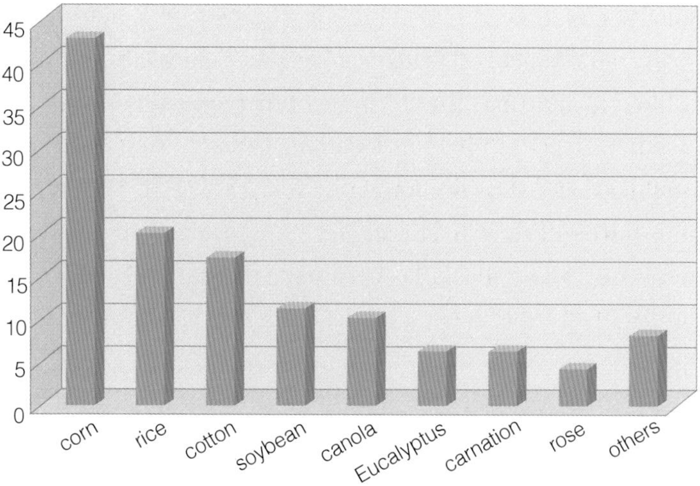

Figure 6-5. Status quo of registered biotech products per crop in Japan

Registration Status Quo

Japan operates the Biosafety Clearance House for the management of biotech crops, and shares information through its website (http://www.bch.biodic.go.jp/english/lmo.html). From 2004 to 2008, the number of LMOs that have received approval for edible feed processing is 125. Corn with 43 breeds, takes up 34% of the whole, and rice with 20 is 16%, followed by cotton with 17 (14%), bean with 11 (9%), and canola with 10 (8%), with these crops taking up 81% of the whole. A notable difference from other countries is that biotech rice is ahead of cotton, bean, and canola.

6-4. Canada

The organization overseeing the environmental release experiments is the Canadian Food Inspection Agency's (CFIA) plant protection department, and application approval information is provided on their website. (http://www.inspection.gc.ca/english/plaveg/

bio/confine.shtml) Environmental safety inspection is done through a process of confined experiments, isolated field trials, unconfined experiments, and then finally the commercialization stage. Confined experiments are held in locations isolated from the external environment, and thus do not need the approval of the competent administrative office. The last stage, commercialization, happens after ample safety assessment has been done through isolated field trial and unconfined experimentation, so it is a relatively simple declaration process.

From 1988 to 2008, 3,579 cases of crops underwent isolated field trial. This is the second most amount of experimentation after the USA out of the OECD nations, and of the 34 crops, the 6 crops of canola (1,448 cases), wheat (440 cases), corn (427 cases), alfalfa (293 cases), bean (234 cases), potato (182 cases) total up to 3,024 cases and take up 84% of the whole (Figure 6-6). Subject genetic characteristics include herbicide resistance, positive sterility transduction, insect-

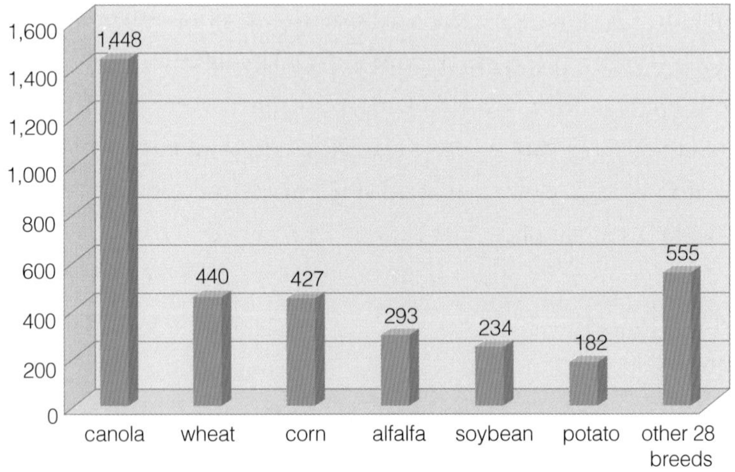

Figure 6-6. Status quo of approved isolated field trials per biotech crop in Canada

resistance, and tolerance against environmental stress.

After isolated field trials, products with outstanding development and market value are selected and have unconfined experiments conducted on them. As of the status quo, biotech plants that have been approved as environmentally safe include 25 cases of corn, 12 cases of canola, 8 cases of wheat, 5 cases of potato, 4 cases of bean, 2 cases of beet, 2 cases of sunflower, 1 case of flax, 1 case of lentil, and 1 case of alfalfa, with a total of 10 species and 61 cases.

6-5. EU

EU member states are supposed to submit a SNIF (Summary Notification Information Format) to the EC's JRC (Joint Research Center), and this data is registered online. (http://gmoinfo.jrc.ec.europa.eu/)

From 1991 to September 2008, a total of 22 nations conducted 2,352 isolated field trials on biotech crops. Nationally, France had the most with 589 cases, Spain 436 cases, Italy 295 cases, England 235 cases, Germany and the Netherlands both with 178 cases, and Belgium 133 cases, with the total number of all seven countries being 2,044 cases, taking up 87% of the total amount (Figure 6-7). The most representative isolated field trials include corn with 775 cases, canola with 379 cases, beet with 252 cases, potato with 274 cases, tomato with 75 cases, cotton with 62 cases, and tobacco 59 cases, with the total number of all seven crops being 1,876, and taking up 82% of the total amount. Modified gene traits include herbicide resistance, pest and virus resistance, and improved product quality (Figure 6-8).

6-6. OECD

The Cartagena Protocol on Biosafety, adopted in January 2000,

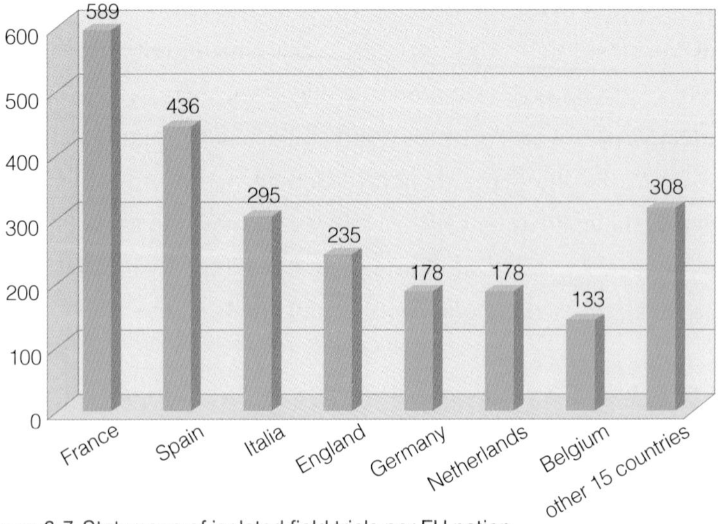

Figure 6-7. Status quo of isolated field trials per EU nation

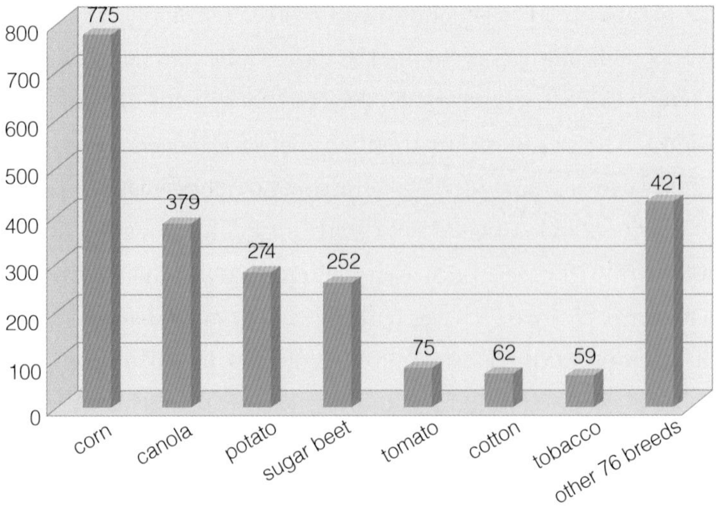

Figure 6-8. Status quo of approved isolated field trials per biotech crop in the EU

announces the regulations on the international transportation of GMOs. However, as there are no international guidelines on risk assessment standards or methods, the OECD, an international organization, established 'The Conference of Working group on Harmonization of Regulatory Oversight in Biotechnology' in 1995. This conference was intended to harmonize differences within countries' risk/safety assessment information and methods, increase mutual understanding, and raise efficiency by eliminating operation redundancy.

At this conference, risk assessment standards, methods, and policies for transformants were written into consensus documents. The OECD BioTrack Online, an information system on biotechnological safety, is being operated. This website includes member nations' data on release experiments, product data, technology and policy trends, and related documents. (http://www.oecd.org/env/ehs/biotrack)

Environmental release experiment data is registered on the BioTrack website, and contains information such as nations that have performed release experiments, release experiment number, year, the organism's popular name, scientific name, biological classification, purpose of the experiment, the inserted gene's traits, approval status, location of release experiments, operating institution and operator, primary government office and operator, and biotechnological category. By the end of 2008, a total of 25 nations carried out 19,902 cases of LMO environment release experiments.

Main isolated field trial nations for transformants are the USA, Canada, France, Spain, Italy, England, Germany, Netherlands, Belgium, and Australia. Of these, the USA takes up 70%, Canada 28%, and France 3%. Of crops, corn took up 38%, canola 11%, bean 8%, and potato 6%.

Biotech products from OECD member nations that have received approval for commercialization are registered online. (http://www2.oecd.org/biotech/default.aspx) The included data is filled out based on documents handed into the OECD by the government institution or development agency. As of today, the total number of approved biotech products is 129, with most crops, 36 cases of corn, 20 cases each of cotton and potato, and 15 cases of canola, with 134 crops registered.

Chapter Seven

Biotech Crops and the Consumer's Option

As biotech crops were introduced to the market the issue of the potential risks on the human body and nature had been raised. Many environmental organizations and consumers have asked the government to provide a regulatory protocols to secure the safety of the biotech products. However, before the establishment of regulatory systems to ensure the safety of biotech crops, the government had already implemented a 'Labeling of Genetically Modified Organism' in April, 2001. Now, the labeling system has settled down that the consumers have chances to take their own choices.

In Asia, Korea was the first to implement the labeling; it was implemented even a month earlier than that of the Japanese. Japan also allows to label genetically modified products only if the government approves of its safety. However, since its approval system was not firmly established yet, Japan did not have a definite labeling system either as were many other foreign countries. Thus, the method of detection or the labeling probe is not an easy process. As the amount of genetically modified products that are sold in Korea's market today is limited, and even in the European Union, the place that first implemented the system in 1998, the system is not being widely utilized. In the European Union, as in Japan, the system was carried out to respect the member nations' right to know the production methods of products that they might consume.

Currently, including Korea, there are the European Union, Japan, Australia, New Zealand, and 20 other countries that enforce the labeling practice alongside with the scientific verification and social verification. Scientific verification is the process of scientific examination through qualitative and quantitative analysis. Social verification is carried out through Identity Preserved Certificates or certificates approved by the producing government which state that the identity of products are preserved to prevent the contamination from genetically modified organisms during the process of the seed purchase, production, preservation, local transport, selection, shipping, and other further procedures. When the genetically modified organism products are imported, they have to be labeled on the Import Declaration and the products are further notified to relevant authorities to follow up during distribution.

The Labeling system for Genetically Modified Organism intends to ensure consumer's right to choose the products by providing its methods of production. However, United States of America that develops and produces the largest amount of genetically modified products, has not implemented the labeling system simply because that could bring prejudice on the safety of products. In Korea, it is anticipated that verification of the safety of the genetically modified organisms would help to settle down the labeling system.

Figure 7-1. Corn chips made from Bt corn and Tomato Puree made from FlavrSavr

7-1. Labeling of Genetically Modified Organism in South Korea

-The subjects of Genetically Modified Organism Labeling

Korea's labeling system on *Genetically Modified Organism* distinguishes between the product itself and the raw materials from which they are produced. It is because the Korea Ministry for Food, Agriculture, Forestry and Fisheries is in charge of the raw materials while the Korea Food & Drug Administration is in charge of the processed foods. Under the 「Genetically Modified Organisms Labeling Instructions」(Korea Ministry for Food, Agriculture, Forestry and Fisheries Notification 2007-43) in the Agricultural Products Quality Control Act the genetically modified agricultural products for sale should be labeled so that consumers can get information about the products. Following this, the National Agricultural Products Quality Management Service has established 「Methods of specimen collection and examination of targeted genetically modified organisms for labeling」(National Agricultural Products Quality Management Service 2001.1.30) and 「Points of labeling the genetically modified organisms」(National Agricultural Products Quality Management Service 2001.2.28).

The Genetically Modified Foods are labeled in accordance with the criteria of the Food Sanitation Act Article 10, paragraph 1, and the notofication of biotech foods (KFDA Notice No. 2004-06 and No. 2007-76) from the 14th of November 2008. Subjected food or food additives (including imported food or food additives) are the one produced by using one or more of the food or food additives approved to import or produce as major raw materials in accordance with Article 15 of the Food Sanitation, and its recombinant DNA or recombinant proteins remain. The list of the subjected food is in accordance with Article 7 of the 「Standards and Specifications of Food」 and 「Health Functional Food Act」, and with Article 14 of the 「Health Food Standards and Specifications」 and is effective from November 14th 2008 as notified. However, food products produced mainly by soybeans, corn, cotton, rapeseed, sugar beet (including soybean sprouts and other vegetable seedlings) as in Sub-Article 27 of Article 3 and additional food products produced from raw materials listed in No 1 to No 27 of Article 28 as major ingredients are defined to be enforced after 6 months from the date of the notification (Food and Drug Administration Notice No. 2007-76).

-Labeling Obligator

The Labeling Obligators for the genetically modified foods in Korea are venders of agricultural biotechnology products, food manufacturers and processors, instant food manufacturers and processors, food additives manufacturers, retail vender, distribution dealer,

Table 7-1. Subjected Products for the Labeling of Genetically Modified Organisms

	Labeling required	Exemted from labeling
Agricultural Products	Genetically modified products approved by the Korean Food and Drugs Administration	Conventional Agricultural Products - Submit Identity Preserved Certificate or Government Approved Certificate *Below 3% is exempted as a threshold level
Manufactured Goods and Health Funotional Foods	Products that are produced and manufactured by using genetically modified crops.	Conventional agricultural products - Submit Identity Preserved Certificate or Government Approved Certificate *Below 3% is exempted as a threshold level of genetically modified crops - In case when the GM component is not in one of the top 5 components -In case when the GM product is not present in the final product *soy sauce, cooking oil, sugars, alcohol (beer, whisky, brandy, liquor, hard liquor, etc)

import trader and functional food manufacturer, trader, distributor and vender defined by the Enforcement Ordinance to Article 2 on functional health food.

-Methods to Label the Genetically Modified Organisms

The labeling of the genetically modified organisms is carried out as follows. It should be labeled by permanent ink, stamping, inscription, so that the consumers can easily distinguish it from the background color of the package of the product or container. Also the size of the letter should be larger than point 10 for the buyers to identify easily the mark 'Genetically Modified Product' or 'Product Containing OO genetically modified product'. If it cannot be confirmed of the presence of the genetically modified component in a product,

Table 7-2. Genetically Modified Organisms Labeling Method

Division	Methods
Label	Main display surface 'Genetically Modified Product' or 'Includes genetically modified OO component' Next to the material list 'Genetically Modified' or 'Genetically Modified OO' *if the genetically modified products' presence cannot be verified, should be marked as 'Might include genetically modified OO'
Method	On the container or cover, the labeling should be done larger than size point 10, inerasable, and be distinguished from its background color - Domestic Products: printed on cover - Imported Products: attached without possibility of falling off (stickers are possible) With no container or cover, should be indicated on a special bulletin
penalties to the infringement	miss the labeling: Suspension of product production for 15 days – a month - 2 months False labeling: Suspension of product production for a month – 2 months – 3 months * In case of inconsistency with the import report, complementary labeling will be issued
	3 years of imprisonment or a fine of 30 million won or less.

the product should have a label as 'might contain genetically modified products' or have a written phrase inside the parentheses as such. Undeclared or misrepresented label will be imposed of upto 3 months of suspension of the manufacturing of the product, 3 years of imprisonment, or a fine of 30 million won or less.

In Korea, the threshold level for the unintended contamination of the genetically modified products is established as lower than 3%.

7-2. Labeling of Genetically Modified Organisms in Other Countries

According to the wide variety of scientific evaluations from Food and Drug Administration, U.S. Department of Agriculture, Environmental Protection Agency, and other trustful organizations, it is generally accepted in United States that agricultural biotechnology products and conventional agricultural products are equally safe. Declaring a strong opposition stance to labelingon the ground that labeling itself is an indirect acknowledgement of the unknown dangers, United States of America sticks to their current voluntary labeling system.

However, as labeling of the genetically modified organism is becoming a big social issue in Korea, Japan and other major grain import countries, the United States government asked the FDA to enact a guideline that distinguished genetically modified organisms from the conventional one; for the USDA, America asked to develop test methods for the genetically modified food. Grain Inspection, Packers and Stockyards Administration of the U.S. Department of Agriculture established a research laboratory to control private agencies to check for the presence of genetically modified organisms in grains exported to foreign countries. Also, it is in charge of certifying the genetically modified organism test kits

Since 1997, the European Union has implemented a mandatory labeling system for genetically modified products. Also, the threshold level for the unintended contamination has been set at 0.9%. Since there is no scientific method to verify the lowest content and exempted items are not established, the system is actually still pending. Recently, the EU Commission strengthened the labeling regulations and included biotechnology feed to its list as well. Even though the original system excluded products that did not have remaining proteins or DNA, the revised version commands no exceptions and orders to label all the genetically modified food.

The United Kingdom has enacted the labeling law since March, 1999. It has strict regulations that include even restaurants to inform customers of the genetically modified product usage. The law strengthened to punish and prohibit further violations.

Switzerland decided to apply the Polymerase Chain Reaction (PCR) system that amplifies the specific gene fragments to verify the genetically modified organism officially. Furthermore, starting from June 2000 the Switzerland government set the threshold level to 1% (Seed 0.5%, feed 3%) for the biotechnological products.

Japan's Ministry of Agriculture, Forestry and Fisheries regulated in notice no. 517 (March 2000) that 24 items of raw agricultural products such as soybeans, corn, potato, rapeseed, cotton, tofu, soybean paste, soybean flour, corn starch, processed foods, etc- will be under compulsory labeling from April, 2001. Moreover, Japan designated its adventitious presence to 5%. For easier promotion of the compulsory labeling, the Ministry of Health, Labor, and Welfare organized the 'Recombinant DNA Technology Application of Food Inspection Methods' and is actively using this. Sampling methods and qualitative and quantitative test methods for biotechnology food are described in detail.

In Australia and New Zealand, the Australia and New Zealand Food Standards Committee (ANZFSC) made the labeling mandatory since December 7th 2001 for the purpose of informing consumers of the genetically modified components of the products that they consume. There are currently six main crops that go through the safety regulation for labeling, which are soybeans, corn, potato, rapeseed, cotton, and sugar beet. However, depending on the future increase of distribution the number of crops subjected to labeling might also increase.

The phrase 'Genetically Modified' is being labeled on the cover or distinctly provided inside the package, when food or raw materials contain recombinant protein or DNA. It is mandatory that this label of the genetically modified food is displayed on the cover. Even in cases of unpackaged materials and piled up fruits and vegetables the labeling is practiced. However, in diners, moving vehicles, or food stalls the labeling is not compulsory. For conventional crops to contain biotech foods the threshold level rate is designated at 1%. More information on how to display and inspect the genetically modified food is described in the user manual.

Based on the Russia Federal Commissioner of Sanitation (September, 1999), organizations, corporates, and individuals who import and produce genetically modified food or medicine must label the products (July, 2000).

There is currently no internationally consensus labeling system or an analysis method. Regarding the international labeling system, the United Nations' Codex Alimentarius Committee on Food Labeling has been dealing with the agenda since 1994 with Canada as its chairman country.

In May 2002, during the 30th Codex meeting on product labeling the different stances between the importing and exporting countries

Table 7-3. Regulatory systems for GM food in major countries

Country / Content	Korea	Japan	European Union	United States
Detectability	Detectable	detectable	Raw materials Used	Undeclared
threshold level	<3%	<5%	<0.9%	-
Subjected items	- Products that have genetically modified organism remnants - In the top five components of the raw materials	- Products that have genetically modified organism remnants - In the top three components, over 5%	All products	-Showing allergic, nutrition difference to the conventional product
follow-up	Documents Inspection and Analysis	Documents Inspection and Analysis	Traceability	-

became the main issue. Since the United States, Canada and other exporting countries disagreed, the international labeling system has not been established. Also, since 2000 a special working group meeting between governments has been convened as Japan being the chairman country. To standardize the system of detection of the genetically modified organism there is a working group chaired by Germany as well.

The 3rd Intergovernmental Special Working group meeting was held in Japan in March 2002, and then in September 2003 the Cartagena Protocol on Biosafety to the Convention on Biological Diversity, an international treaty, took effect. According to this protocol article 18 clause 2, genetically modified products for seed, edible, and experimental use should be labeled.

7-3. Authorized Analysis Methods

The two commonly used genetically modified organism detection methods are Enzyme Linked ImmunoSorbent Assay (ELISA) and the Polymerase Chain Reaction (PCR). Also, for biotech crops, herbicide treatment detection methods can be applied.

a. Enzyme Linked ImmunoSorbent Assay Method

Enzyme linked immunosorbent assay is a method that detects the protein in a immunological way to recognize the genetically modified organism. This method is suitable for detecting targeted proteins from the biotech crops or the raw materials. Biotech crops and the conventional crops have different expression levels of protein depending on the tissue or organ, so when the transplanted gene's protein does not express itself in the tissue or organ the detection becomes quite difficult.

There are two different ELISA method to detect the biotech crops; one is the Lateral Flow Strip Technology* to test the presence of GMO qualitatively and another is the Microtiter Well Assay method to detect GMO quantitatively

The statistical precision of the common ELISA method at the reaction is less than 5%, and the concentration derived from the analysis is less than 10% which shows a variance of less than 15%. Therefore, 1% of biotechnological products could be analyzed to be 0.8 to 1,2% range, which needs further study on the error range of the threshold level by quantitative analysis.

In recent years, the American Association of Cereal Chemists (AACC) held the analysis trial to validate Mon810 developed by Monsanto, a pest resistant corn, through the Institute for Reference

---

* Lateral Flow Strip Technology: A diagnosis method in which an antibody specific to the recombinant protein produced by a gene newly introduced into a GMO is combined to a test strip to trigger antigen-antibody reaction. It is used as a qualitative analysis technique to identify whether an organism is GMO/non-GMO.

Materials and Measurements (IRMM) in Belgium. The international ELISA method validation was carried out with 0, 0.5, 1, and 2% samples prepared by the Certified Reference Material (CFM) and 0.3, 0.75, 1, 1.25% samples prepared by the Canadian POS Saskatoon were analyzed by 40 labs in 20 countries. , The National Institute of Agricultural Science and Technology and the Food and Drug Administration of Korea also participated in this validationerification. From all the efforts, the results were shown as such; accuracy of ELISA 96.7~100%, repeatability 6.56~18.4% (relative standard deviation), and reproducibility 13.8~23.5% (RSD).

b. Polymerase Chain Reaction (PCR) Methods

Another method that is being widely used for the detection of the biotechnological crops is the polymerase chain reaction (PCR). The PCR technology was first proposed theoretically in 1971 by Kleppe and Khorana (J Mol Biol 56: 341) and then applied in experiments from 1985 by Mullis et al (Science 230: 1350). The PCR technology caused a revolution in the field of molecular biology and biomedical applications. Thanks to the chemical property of DNA with high temperature tolerance, sensitivity of PCR , and the simplicity of technology, the knowledge and experience for PCR have been well accumulated. Moreover, with the automation of the PCR technology, most of the detection methods now include the PCR.

The PCR detection method can amplify a specific gene region or a fragmented gene, which allows the detection from the raw materials and processed products to even 0.01% level. When applying the PCR method to detect GM products, genetic information on the transformed recombinant genes is indispensible. This is a problem because the genetic information cannot be obtained without the voluntary provision from the GMO's developing corporates. Moreover, even for the same gene, for example the pest resistant CryIA(b), since different companies have slightly modified nucleotide sequences differently, the development of the primer itself is challenging.

Therefore, sharing amongst domestic and overseas research centers of information and building up databases based on the genetic information are necessary and crucial. Typically, the most popular part in the PCR test is the primer of a cauliflower mosaic virus (CaMV) 35S promoter. This acts as an initial screen for the detection of genetically modified organisms. Though it is not a genetically modified organism but this virus exists in the natural environment, it can show a positive reaction. Hence, it is important to develop a specific primer to determine the precise identity.

In PCR detection one has to be careful with cross-contamination between samples, false negative responses due to the contaminated PCR remnants, primer specificity, detection sensitivity, reproducibility and accuracy, and contamination of the inhibitory substances of the PCR amplification materials during the DNA extraction or a negative response due to the DNA's degradation,.

An example instance of the PCR validation is the Ring Study, with unique primers for 35S promoter and the Nos terminator carried out by 13 EU member nations through the Joint Research Center (JRC) in Europe, and 29 labs. The herbicide resistant Roundup Ready Soybean (RRS) from Monsanto and the pest resistant corn Mazimizer (Bt176) from Novartis were the samples used in the cooperative research. These samples were provided by the IRMM, and the 16 samples were each mixed by 0, 0.1, 0.5, 2% of biotech samples. The unique 35s promoter primer detected RRS from 0.5% through the PCR screen, and Bt176 from 2% at 100% frequency. The unique primer for Nos terminator showed a 100% accuracy when RRS was above 2%.

Until now, many countries including the European Union regulate the labeling system for genetically modified products by law. Since there has been no internationally approved test method required by law, each country has its own methods. In the case of the European Union, member governments are using the approved methods by the European Commission provided in CRL (http://gmo-crl.jrc.ec.europa.eu/statusofdoss.htm).

Germany has an official detection methods for the transgenic potato, transgenic microorganisms for fermented sausages, yogurt, and other products. These detection methods are officially mentioned in the 35th article of the Food Act of 1997. In this law, it requires the

Table 7-4. Run-Time and the pros and cons of analysis methods

Analytical Methods	Run-Time	Pros and Cons
Qualitative Analysis Lateral Strip Flow Test Kit	(Total:20min) Sample Grinding : 10 min Strip Test : 10 min	(Pros) Fast detection time (Cons) Low Accuracy Impossible Quantification
PCR Using Unique Primer	(Total:13hours) Sample Grinding : 30 min DNA Separation : 3hrs 30 min DNA Purification and Quantification : 3hrs PCR : 2hrs Electrophoresis : 3hrs Result Analysis : 1hr	(Pros) Qualitative Detection of GM crops (Cons) Impossible Quantification Takes more time than the Strip Test
Quantitative Analysis Real-time PCR	(Total:18hours 30min) Sample Grinding : 30 min DNA Separation : 4hrs DNA Purification and Quantification : 3hrs RT-PCR Reaction Preparation : 2hrs RT-PCR : 5hrs Result Analysis : 4hrs	(Pros) Quantitative analysis of GM agricultural products Can be quantified without standard materials (Cons) Takes a lot of time Requires technical skills Requires expensive equipment and reagents

test results conducted jointly in university laboratories, private research institutes, and joint government agencies to be included.

To label imported biotech crops in Korea and the European Union, it is mandatory to provide the nucleotide sequence of the transformed gene, reference samples, related antibodies and antigens to test recombinant proteins by the company that produced the products.

Though Switzerland is not the members of the EU, it has the genetically modified organism labeling system based on EU's. Thus, any products containing more than 1% of the genetically modified organism must be labeled. In the United Kingdom, the Central Science Laboratory of the Ministry of Agriculture, Fisheries, and Food has been designated as the UK Accreditation Service to detect genetically modified products. The Institute for Reference Materials and Measurements (IRMM), one of EU's Joint Research Center, produces reference samples for biotechnological soybean and corn.

In Korea, the Korea Food and Drug Administration bio-food team and the National Agricultural Products Quality Management Service of the Ministry for Food, Agriculture, Forestry, and Fisheries developed detection systems for soybeans, soybean sprouts, corn, canola, and cotton through quantitative and qualitative analysis. These systems are being officially used for the detection of the GM products or the processed goods.

The most common test to detect genetically modified organisms is the polymerase chain reaction (PCR). This detection test is a method that amplifies specific fragments of the genetically modified genes or the regulatory region of the gene. Thus, even with a very small amount of the DNA sample, through PCR the sample can be amplified to a large amount in a short time to examine the properties. When this method is applied, it is possible to detect the transgenic materials in even raw materials such as rice, soybeans, and processed foods such as tofu and soy paste.

Recently, a detection method through antibodies that react to the GM proteins are in the market as a portable test kits. This handy protein detection kit can detect herbicide tolerant soybeans and Bt protein containing corn with no difficulty. Also, this method can be easily applied in grain storehouse or in the field in very short periods of time.

Whether in a DNA method or a protein method, the most important issue is to develop an internationally acceptable scientific and statistics based sampling system. For the accurate detection of the genetically modified organism, the procurement of the reference samples is very important. These samples include genetic sequences and antibodies of all the newly developed genetically modified organism. Procurement of genetic sequences and samples are difficult without the voluntary provision of the corporates, hence, creating a data base and an information sharing system is needed.

With the mandatory labeling of genetically modified organism, importers and distributors get ready its own inspection center to test the presence of the GM products or add the test function to the currently existing research centers or private companies. Government-funded domestic institutions are the Korea Food Research Institute, the Korea Consumer Agency, Korea Health Industry Development Institute, and Korea Merchandise Testing and Research Institute. There are also private companies such as the Korea Gene Analysis Center, Kogene Biotech, GD Biotech, and Nexgen, etc. In such corporates, quantitative and qualitative PCR equipments are prepared for the consumers demand for the detection of genetically modified organisms and also for the sale of the transgenic detection kits.

Despite the original purpose of the labeling system to satisfy the consumers' right to know on the process, many people misunderstand the meaning as a label for unsafe. Even though there are many concerning about the potential risks of the genetically modified agricultural products, there are no valid scientific evidences for the danger of the GM products. Hence, when considering the scientific evidences, many scientists agree that approved genetically modified agricultural products in the market are safe. Moreover, the widely commercialized genetically modified crops by developed countries already went through safety test which verified their safeness. Therefore, as the genetically modified organism's safety being the top priority, the government and consumers should try to have a balanced perspective for the further advantageous applications of the crops.

Biotechnology on a Table

Chapter Eight

Disputes over and the Scientific Facts of Biotechnological Crops

As biotech crops are cultivated by farmers, they are bringing great positive effects economically, but some people are raising their voices regarding safety issues. Some people and organizations claim that both the development of genetically modified organisms (GMO) and the harvest of Biotech Crops should be prohibited because biotech crops could have negative effects on human body when digested in forms of food Health Safety (Health Safety) or biotech crops could cause the appearance of unwanted new species and unkillable super-weeds when the artificially injected DNA moves to the ecosystem (Environmental Risks). How much of this information is scientifically proven? Let us take a look at the scientific truth, mainly the facts that have reported biotech crops to be detrimental to humans.

8-1. The Safety of the Human Body and Related Scientific Facts

As GMO Technology rapidly developed and GMO Food have been placed on dinner tables, controversies have been raised on whether GMO Food might have negative effects on health. Let us analyze the scientific backgrounds of these controversial reports.

a. Doctor Pusetie's Research

On August 10, 1998, Doctor Pusetie, from Rowett Institute of Scotland criticized the safety of genetically modified foods on TV, claiming that a rat that ate transgenic potatoes with *lectin* gene from jack soybean had its immune system such as liver, spleen, and thymus greatly damaged. After he had given transgenic potatoes to five rats for 110 days, (about ten in human years), they suffered slight underdevelopment and immunosupression. However, other rats that ate transgenic potatoes with anemone-related genes had not shown such results.

As disputes aroused, Rowett Institute organized an Audit Committee to verify Doctor Pusetie's Research results and announced that his conclusion was wrong. After experts examined thoroughly, they found out that in terms of the assertion that said there were differences in organ weights of the experimental rats, there were no indications of change. Furthermore, when experts compared the white rats that were fed with transgenic potatoes with *lectin* genes and other white rats that were fed with normal potatoes with lectins added on them, they could not detect any growth difference between two rat groups or statistical significant difference in growth rate. Moreover, it was difficult to admit both biological and statistical differences of the immune systems because there were not enough tests. Although the transgenic potatoes with *lectin* that prevents nematode attacks, may be assumed to cause harmful effects to rats, the actual result revealed no such effects. After all, Doctor Pusetie had to leave the Rowett Institute for announcing such defective research without any permission.

On June of 1999, The Royal Society also examined some related data and judged that Doctor Pusetie's conclusion could not be driven out due to his various mistakes. Doctor Pusetie sent messages to the Royal Society that it was not proper to testify the results outside the lab since it was an inner report in the lab, even though he publicized his report.

Then Doctor Pusetie once again released another research paper claiming that transgenic potatoes with *lectin* genes affects rats' organs and immune systems, on October 1999, in a journal called Lancet. However, many scientists posed some questions about whether the result of his report was appropriate. In terms of this, the editor of Lancet released his report on condition that the report had several imperfections in its planning and analysis. However, that publication was to discuss biotech crops between the scientists, public, and the mass media.

In that experiment, six types of feed potatoes; natural potato, potato including *lectin*, and transgenic potato, which were all prepared in raw and fried forms were given to rats. Here, the effects to some of the organs and immune systems of rats caused by the transgenic potato were questioned. Some have opposed that the effect could be driven due to the stress from protein-lacking food, the type of potato, or the low digestion rate of food.

The biotech potato used in Doctor Pusetie's Experiments has never been commercialized, but only been used in researches. The *lectin* also exists in soybeans and it is harmful to humans since it lacks in nutrition, so the effort to reduce this effect has been continued. Still not gaining its recognition, this event is the representative case that criticized the safety problems by testing the transgenic crop which humans do not consume.

This was the first case to start the dispute about the safety problems of biotech crops and it is still being cited frequently. It is also represents how much of an impact a misreport has on society.

b. The StarLink Corn Case

In 2001, the transgenic corn StarLink was imported to Korea and Japan, but it was sent back. It was approved as feed, but not as food since it had a transplanted gene, which made insecticidal *Bt* protein be digested slowly by the enzyme pepsin in the human body. Normal protein would take 2 minutes to be digested but the insecticidal protein would take about 3 minutes, so it was only allowed to be used as feed just in case it causes some allergy to humans. However, during the distribution, some of it was mixed into the edible corn and was used in Mexican-Style restaurants to make taco shells, which caused a social uproar. Though it was a managerial problem that allowed some of the transgenic corn which had been scientifically examined and proven to be safe to be mixed into the normal corn, this case was misinterpreted as though it was caused because GMO corn were harmful to humans. This case is like the big uproar of how the low quality of beef tallow, originally for industrial use, was used in making Ramen. This event was caused by transportation problems and had nothing to do with the safety of biotech cops. Rather, the safety examination gained more reliability by classifying the transgenic corn as feed.

c. Weevil in Australia

In 2005, a research team of CSIRO, a research agency of Australia, inserted normal bean's α-amylase inhibitor gene into peas, in order to repel weevils. The gene is only found in beans, but not in peas and by the manifested α-amylase inhibitor, the weevil suffers indigestion so before it even harms the pea, it would starve and die.

The animal testing to examine the transgenic pea revealed that the rats showed an immune reaction. Then the CSIRO had to stop the research in 2005 due to the pea's immune reaction. The transgenic pea had only been on research and development phases, so it has never been commercially harvested with the Australian Government's permission. Therefore, no one had consumed it. This case aptly shows how the biotech products with safety problems cannot be commercialized.

There was also a case of the Brazil nut. Soybeans lack in nutrition, since they only have small amounts of essential amino acids with sulfur. So to use soybeans as feed, the essential amino acids had to be supplemented. The world famous American Seed Institution, Pioneer Hybrid transplanted the *2S albumin* gene to make transgenic soybeans with improved essential amino acid, methionine and cystine in 1996.

However, as it was confirmed that the allergy of Brazil Nut comes from the protein the gene makes, the Pioneer Hybrid had to stop the research, wasting 100 thousand dollars. Therefore, Brazil Nuts from the transgenic soybean had no chance of causing allergies, unlike the natural Brazil Nuts.

d. Irina Ermakova

On October of 2005, Doctor Irina Ermakova from the Russian Science Academy of nerve reaction and nerve physiology stated in Russian nongovernmental organization's symposium about recombinant of genes, that from the experiments, rats eating biotech soybeans had six times of higher death rate in three weeks after the birth and showed lower body weight than the rats eating normal soybeans, and the England mass media reported it.

The Advisory Committee on Novel Foods and Processes (ACNFP) of United Kingdom FSA announced statements that criticized Doctor Ermakova's research, on December, 2005. Likewise, his report was not performed as an academic essay and it was unreliable, since it was not confirmed by any other experts of the field. However, in 2004, an essay was officially announced that from an experiment through four generations of rats, biotech soybeans affect neither the death rate nor the growth of rats.

e. The Mass Death of Sheep in India

KBS broadcasted how sheep and goats that consumed biotech raw cotton in India died mysterious deaths. If the biotech cotton affects the sheep and goats, then within 10 years, the teenagers eating them would probably die too. If a kid about 10 years old continues eating cooking oil, it is likely that the disease would occur in 10 years.

What KBS reported was that the sheep and goats in Andhra Pradesh of India had eaten genetically modified *Bt* raw cotton which is resistant to insects, and then died. However, as shown in case B, the *Bt* protein can affect certain insects, but not sheep or goats. If the sheep and goats died because of the biotech cotton, sheep and goats in other regions growing the biotech cotton would have died, too. However, there were no similar reports from China, Australia, Brazil, Argentine, USA or any other regions of India where they grow biotech cotton.

Indian regulation institution, Genetic Engineering Approval Committee said "No report or analysis can conclude that the mass deaths of sheep and goats in India were due to the Biotech Cotton." Moreover, the Principal Farm Secretary, CVSK Sharama in 4th of April, 2007 clarified to the public, "Scientists already checked that

Figure 8-1. Commemorative photograph with herdsman in India

the mass deaths of sheep and goats were not because of the biotech cotton, so cultivating it is not inhibited in Andhra Pradesh." Left-over agricultural pesticides and excessive amount of nitrate were mentioned as the cause of the deaths. It is not specifically defined. In India, the transgenic cotton was harvested in the area of 6.2 million ha (66% of the total cultivation area of cotton). If it was so dangerous, it would not have been cultivated in such a large area.

On June of 2008, a group formed by CUK (the executive of Consumers Union of Korea), reporter of Joongang Newpaper and biologists visited the local cattleman of Warangal city in Andhra Pradesh to verify the truth. They said there was no such mass death and sheep preferred the biotech cotton, since it was clean and not sprayed of agricultural chemicals. They also mentioned how sheep and goats do not eat shriveled yellow cotton.

8-2. Safety within the Environment and Related Scientific Facts

Among several opinions, these are the main points disputed about the effects of the Biotech organisms on the environment. It is possible that the gene moves into surrounding wild plants and so unkillable superweeds might emerge unexpectedly. The incidents leading to statements that the genetically modified organisms can affect the ecosystem have not widely attracted social attention, so they are not so much reported in the mass media. However, let's find out some related scientific facts focusing on those that are mentioned.

a. Emergence of Super Weeds

There are some worries that GM crops survive better than normal crops. Therefore, in the future only the GM crops would stay, leading to the emergence of unkillable superweeds. To verify the validity of the opinion, in 2001 on the academic journal Nature, English Scientists tested with four GM crops for 10 years and proved there was no supporting evidence for the statement. They had harvested GM potato, corn, rapeseed and sugarcane with herbicide-tolerant gene and insect resistant gene, in twelve different regions. By observing those crops, scientists could not find any different wintering ability and survival ability or the existence of a superweed. Compared to the wild crops, GM agricultural products were not stronger, so within 4 years they disappeared. This research made England and any other European countries which strongly opposed harvesting GM crops and trading with countries that harvest them in confused. Neither normal crops nor the GM crops can survive out of the care and protection of humans.

There is also another claim that resistant pollen (gene) in a GM crop might move to another normal plant to produce herbicide-tolerant weeds. Generally, when harvesting some crops, it is out of interest whether the gene moves into nearby crops or not but considering the genetic side, it is possible that the gene moves into another surrounding crop. However, accidental crossbreeding of wild and GM crops is expected to be hard, considering how the success rate of crossbreeding different types of plants is very low in traditional harvesting. Even if that happens, without spraying herbicide, only herbicide-tolerant crops surviving and others all dying is not possible.

This statement also lacks scientific prove. Each gene having tolerance to herbicide can only resist to each single factor of the herbicide, so even if the superweed having resistant to a herbicide emerges or surrounds the GM crops, spraying different type of herbicide can kill it. There is no need to worry about the appearance of a superweed, if growers stop using the certain type of herbicide. The claim that diversity of organisms can be damaged by genes moving is especially highlighted in original country of the crop or in the countries with various genes. However, generally crops cannot survive without the protection of humans and the farm land where crop growth is thoroughly controlled. This is different from nature crop growth. Does anyone see rice or peas growing themselves in the mountains and fields, not in paddy or farm?

b. The Killing of Monarch Butterflies Caterpillar

The insect-resistant corn and cotton are invested and harvested by a toxin gene called *Bt* (*Bt* gene) separated from microorganism (*Bacillus thuringiensis*). There was a claim that the insect-resistant crops in paddy can accidently harm non-target organisms and some research were presented to support the claim. Digestive organs of Lepidoptera insects are alkaline or neutral and the protein of *Bt* gene interrupts their osmoregultaion of epithelial cells of intestine and eventually kills them; but, it is no harm to humans whose intestine is strongly acid, so it has been permitted to be used and sold as biotic pesticide.

However, as the fields harvesting insect-resistant corn increased in USA in order to control the corm vermin Ostrinia furnacalis (Guenee), argument that the caterpillars of Monarch Butterflies which eats milkweed near the field can accidently be harmed.

The first research was performed in Cornell University, but it was carried out in laboratory that was inappropriate for caterpillars of Monarch Butterfly to grow and they were fed corn flower pollens. The research caused even more controversy and the professor who fulfilled the research admitted that the experiment lacked conditions.

Thus in 2001, six teams from America and Canada performed a precise research by cultivating packages on insect-resistant corns and reported their result to the National Academy of Science, USA, that such claim was not true. Researchers argued that pollens of the insect-resistant corn lacked the *Bt* gene; before the caterpillar eats 4,000 pollens, its growth and development is not influenced; the corn pollens that could possibly be stacked up per of its food, mild weed, are maximum of 120, so caterpillars could not possibly be harmed. They also concerned the distance between the corn field and the mild weed, the overlapping period when the corn pollens spread and the caterpillars grow and reported that the corn showed no harmful result to caterpillars of monarch butterfly. According to their research, Environmental Protection Agency (EPA) announced that harvesting insect-resistant corns with *Bt* gene does not harm the growth and development of other insects like monarch butterflies and that there would be no extinctions of such insects.

Meanwhile, a seed company that has invested crops resistant to pests like Ostrinia furnacalis with large amounts of money would have to face big problems. If they harvested only crops resistant to Ostrinia furnacalis, it would grow resistant to *Bt* gene as generations pass, and then eventually the all of the natural Ostrinia furnacalis would have resistant to *Bt* gene. In 1996, the EPA insisted on making a refuge of normal corn when harvesting insect-resistant biotech corn. In other words, the area of the refuge should be 20% of the area of insect-resistant corn, when harvesting with pesticide, four% when harvesting

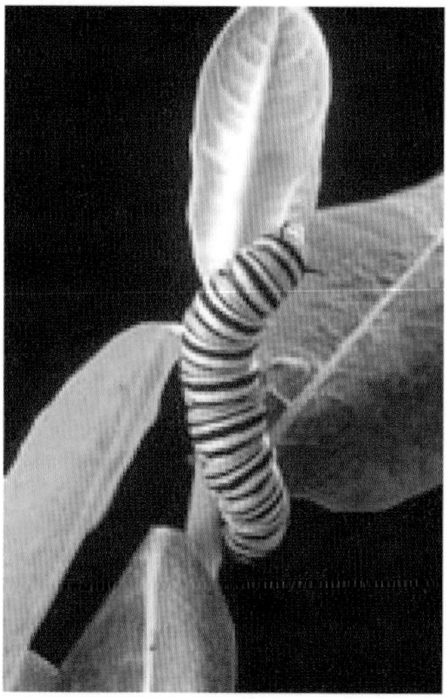

Figure 8-2. Caterpillars of monarch butterfly

without pesticide to induce the mating of the pest with resistance and without resistance and to slow down the rate of the pest having resistance. The reason for this action is not certain whether it is for the Ostrinia furnacalis or the seed company.

8-3. Conclusion

Currently, the information we received is great in diversity and quantity. However, it is difficult for us to lucidly comprehend or ensure the validity of the information. The common inclination is to consider facts about society or politics which are aptly fallacious or biased. What if this information is about science? People take in scientific information without doubting its authenticity or the range of validity. This is because it is science. Science is always veracious, it is factual, and cannot lie. It is true in principal. However, what happens when one alleges unverified facts with scientific logic? Most people would have to believe it, especially when it has to do with the food that we eat. We traditionally are slightly generous to consider inevitable circumstances about breaking societal rules, but stringent in the so-called meddling with food.

The era of biotechnology that artificially manufactures DNA to transplant it in organisms for cultivation or process it in incubators to utilize as food is about to become real. Biotech Food is already on our families' tables. The dispute about Biotech Crop has been heated for the last 10 years, and has been slightly mitigated, since the international Biosafety Protocol has been inaugurated. However, the interest in Biotech Food has not cooled down. This fervor is seminal in observing and restricting science in order to develop it only for the sake of humanity. True observation, however, should be based on scientific facts; when it is not, consumers receive faulty information. This phenomenon is common in only biotech organisms.

Chapter Nine

The Possibility and the Future of Genetically Crops

According to the United Nations, the world population would constantly increase and eventually be 9 billion in 2050. Thus along with the population growth, the world demand for food must also increase. Until now, we have expanded the cultivated acreage, used chemical fertilizer and pesticides and also have developed and harvested a high-yield variety like Tong-Il grain in order to increase food production. However, along with constant industrialization, the area of the cultivable field decreases and the usage of chemical fertilizer and pesticides are limited for the environment protection, so that these kinds of solution have its limits on increasing the food supply. Moreover, the demand of consumers has been diversified, so the importance and the necessity of food resources' breeding improvements have been increased. Therefore, scientists have used biotechnology in order to effectively develop new breeds.

To overcome some limits within the traditional cross-breeding, trait transformation with useful genes is evaluated as the most effective way. The 21st century will be the era of customized crops. Most of the GM crops harvested nowadays are either resistant to pests or herbicide, that they decrease the production cost or increase the production.

"The First generation of biotech crops" is classified by its transformed trait. Those crops were given new agricultural features that were impossible to gain from the original cross-breeding method. They have the profitable characteristics, such as resistant to herbicide or tolerance to pests, for producers like the Seed and Pesticide Company and farmers.

"The Second generation of Biotech Crops" includes crops such as soybean or canola oil which changed the process of making fatty acid and tomato with extended expiration dates. These products have improved manufactured traits or reduced the production costs, and are now just ready to be sold in the markets. They are advantageous to the distribution or food processing industry.

"The Third generation of biotech crops" is called functional foods, like the Golden rice with strengthened vitamin A, since those crops have intensified nutrition or medical factors, such as edible vaccine, anticancer substance and depressant. These are also called the biotech crops for next generation because they are useful to and found by customers who have the purchasing power.

Along with the second and third generation of Biotech Crops, a new type of food that has increased its added value by improving certain nutrition and health functionality would constantly be developed and propagated. Technology of transplanting gene would be advanced even more that universal, effective safe and stable way would be invented. After all, procuring the useful gene to define certain trait would determine the goal achievement and industrial competitiveness.

However, it is still hard to define a certain gene that determines an organism's traits. The genetic information of Arabidopsis thaliana and grain are revealed, but only functions of 5% of the genes are known. Gradually through complete decoding of genes and innovative biotechnology, it is overlooked that new era would effectively inserted, set excellent genes and produce mass amount of superb kinds, as if replacing machine component. Accordingly, safety problems within environment or food would meet with the demand of customers by precise machine and the development of technology.

These are some few special crops under development; tomato with increased amount of carotene and delayed ripening property; tea and coffee without caffeine; coffee that can be temporally harvested; blue cottons for jeans; tobacco without nicotine; poppy without morphine to extract cooking oil preventing heart disease; vaccine tomato for virus of child pneumonia and bronchitis; orange tree with shortened generation to harvest early; insect-resistant crops like aphid by natural substance, and so on. The technology that transplants and expresses various genes by forms of mass DNA at the same time would be generalized not just determining monogenic feature.

The biotechnology industry of 21st century has many domestic and foreign researches and investments. However, basic progress and more development cannot be expected without having right awareness about the safety of biotechnology. This not only applies for biology, but also other fields of science. In developed countries, crops that are manufactured by DNA recombination are managed and evaluated with the safety systems from the basic research phase, and then circulated in the markets. In Korea, there are no crops yet commercialized, but the guideline of the research phase is not settled. Therefore, new understandings about the safety from researchers are demanded to henceforth secure of safety. Moreover, Korea should also try not to cause social anxiety from those safety installations.

From 1994, when industrialization started and 13 years after, more than 38% of soybeans, corn, rape seeds and cotton are worldly harvested by biotech tools. In the beginning, they were mainly resistant to pests and herbicides. However, in the next generation, the new types of genetically manufactured crops with intensified certain nutrition or health functions would be constantly developed and commercialized, not simply having resistance. Along this progress, mass securing of characteristic genes is the most necessary for its directly related to industry competitiveness. DNA recombination technology is composed of finding and recombining useful genes; reproducing as a complete organism by transplanting and redifferentiating genes to plant cells; and finally, commercializing the products. The industrial competitiveness of transgenic technology is determined by the effective excavation of useful genes and effective transforming technology that transplants recombined gene. Thus, developed countries promote biotechnology as their national strategic industry by focusing on and supporting the excavation of genes through nations. As our self-sufficiency rate of grain is only 26%, we should try hard to increase our agricultural productivity and export the seed or technology by developing the biotechnology.

Meanwhile, there have been some concerns about genetically manufactured crops safety problems on food or environment. Some disputes about producing and consuming those crops. So, each government provided some institutional strategies to secure their objective safety. As time passed, the advantages of genetically manufactured crops stood out and their harvesting area increased, consumers tended to show better understanding towards them. Particularly, Europe which had been prohibited importing transgenic crops stopped the prohibition. Afterwards, the actively increased the harvesting area from 2005 and the institutions tried hard to develop products. In the begging, the biotechnology had shown poor progress unexpectedly, but it must be the era of biotechnology in the 21st century.

After the procedure of producing high functional crops, Biotech Crops would be used as bioreactors that effectively produces mass amount of high quality medicine components, or used as methods to produce useful substances like biodegradable plastics and energy resource factor. If so, the potential of biotechnology is remarkable. Then, it would be settled as a practical technology that can benefit the majority of people.

Translator's Note

As a student, I've always sought for my true interest and aptitude, and for what I should do in the future. In that period in my life, when I consistently contemplated about my future career, I attended the lecture "Biotechnology on a Table" by Professor Choi, Yang Do. The lecture was based on a book with the same title. Listening to the lecture, I thought that it was the first time in my life that I was totally mesmerized by a field of study.

The lecture provided general information about recombinant DNA technology and its application to the real world. I was astonished at how a transformation in the most basic units of life, like DNA, can result in a significant influence in an organism's traits. What was even more fascinating was the real world application. Biotechnology was more than just a field of study. By making the best use of useful traits by recombinant DNA technology, modern biotechnology provides the ultimate key to solving fundamental causes of previously unsolved world problems like poverty, diseases, miserable aftermath of warfare.

Although humanity has achieved tremendous development in terms of civilization and technologies over the past centuries, the rapid developments have also given rise to negative consequences like the creation of the third worlds followed by wealth polarization, pollution, and warfare. Now it is the time to develop technologies to cure and resolve such chronic problems, and to me, it seemed biotechnology provides a large step to accomplishing the common goal of the prospective generations.

At the same time, I thought that this useful information should be shared with more people. Not only the scientists and the ones engaged in the field of study, but everyone who is the component and the owner of this world should learn the knowledge that could make a step toward solving the problems of the world that we share. That is why I decided to translate this book into English with a few of my high school friends who share the same interest. Translating this book into English, I hoped that people from all over the world could understand what biotechnology is to humanity today, and take more interest in it.

I give special thanks to Professor Choi Yang Do, the author of the original version of the book "Biotechnology on a Table", for giving me the honor to translate his book, to all members of "Transbio" who put their best endeavors in the translation of this book, to "Purungilbook" that allowed the publication of this book to the world, and to everyone who helped me in the process of publication of this book.

Kim, Nawoo

***

Before high school, I always thought I would end up as a linguist, editor, or reporter. Anything science-related was far from my interests. However, in the 9th grade, I had the chance to study AP Biology. It was a life-changing experience, because I realized for the first time what a relatable subject Bio actually was.

I believe being so relatable is the true charm of the field of Biology. Through this book, I hope English readers across the world will learn how biotechnology is closely intertwined with our own lifestyles (including the dinner table!), and gain interest in the various social issues that arise due to advances in science. Happy reading, and good luck!

Choi, Joo Hee

***

As a kid who was born with atopy dermititis, a direct result of the pollution caused by misguided technology, I've always been interested in trying to make a world a cleaner, healthier place and get rid of the taints in the environment caused by humans. But as I learned about the teachnology behind the pollution, instead of condemning it I've learned to embrace it and try to make it better. We can't fight science developing in order to make human lives more comfortable but we cannot let it ruin our lives by not thinking about he consequences first. Then the moral debate about the GMO caught my attention

Science is a work of percision and cold, hard facts but is not in a vaccum space human imperfections cannot touch either. The more I become fascinated with the study of biotech and the wonders it can do for humans the more I'm concerned how science can be misguided to harm rather than help. Or in some cases, be driven by fear and prejudice and reject potential life saving technology. By taking part in this project I wanted to prove that despite its social stigmas, GMO can be the future of the food industry and not harm the environment. I also wanted this to be a learning experience for future aspiring scientists to never stop trying to innovate, change and look for room for improvement. While GMO admittedly have some problems in the past, the insessant work to try to make that change was what made GMO the desirable technology that is being used today. I hope whoever was not initially interested in this field, or had the wrong idea of biotech and what it can do for us changed their minds after reading this book.

Han, Hae In

***

Hi, my name is JiYun Han and I am a senior at Hankuk Academy of Foreign Studies. As I was translating this book during the winter vacation, I regained my interest not only in English but also in Biology. Since I was a 5th grader in elementary school, I wanted to pursue my future career in the field of Biology or Dental Studies. I originally wanted to be a doctor (especially a dentist) after I watched a documentary film about the life of Dr. Albert Schweitzer and his love and help towards the people in Africa who could not receive any treatment since they did not have enough money. As I got into middle school and high school, I became more attentive to Biology itself and liked to search for the basic principles that things around us work under. Thus, I did a lot of Biology related activities especially in high school. I participated in biology club activities and the experiments. When translation of this book was suggested, I was really excited since I thought of this as an opportunity to enlarge my interest in Biology and a chance to show my English skills. At first, it was not as easy as I thought it would be. I just thought that it would be no different from simply translating Korean to English. However, it was very different in that it was not simply translating. I had to understand all the materials that are mentioned in the book and find away to effectively convey the meaning of the sentences. Through the translation, I obtained more chance to spend time on learning Biology and developing my English abilities.

Han, JiYun

***

Eight years old, I was crying in front of a photo of an African child suffering in hunger and pain. I don't remember the exact photo, I don't remember the exact reason why I cried but I remember the ache that I felt in my heart that made me determined to help the children in the world to stop suffering.

I was never aware that I would choose biotechnology to help children but I have and I did.

Biology and technology combined equals biotechnology. Before coming to high school I wasn't even aware of that and I also wasn't aware that I would be choosing this field for my path. The only thing that I noticed was that science has always been around me because I have lived with a scientist father. When I came to high school, though, I realized that science was not just a part of my environment for the past 18 years but something that will continue to be a part of my life as I chose that path to help the children around the world who are suffering. With this decision I have since then continued to focus more on biotechnology. One day I hope to truly become someone who is helping the suffering children by wiping their tears and helping them smile.

Translating this book has been another interesting journey for me. Unlike the previous experience of translating a Korean folk tale, this book was challenging with scientific concepts required to be translated in the scientific term. However, the fact that I have learned a lot of information regarding genetically modified organisms has certainly been helpful. I really hope that those who read this book would be able to have a better nonbiased view of GMOs (Genetically Modified Organisms) and find science an interesting subject that is really close to our lives.

Kim, Gaon

***

This isn't merely a guidebook. <식탁 위의 생명 공학> was not conceived in a purely academic tone. Instead, the story of biotechnology of the prestigious professors and researchers was one for those who have wanted to understand why we should take care of the mechanisms of genetically modified organisms. Some could possibly find this difficult to read but each chapter has specific historical accounts and data that will help readers to achieve their goals. It is altogether profitable, and I can see in the course of the development of science the outlines of intellectual and spiritual evolution.

I was delighted, also intimidated, to be invited by my peers to try my hand at this masterwork. I have long wanted to do work regarding biosafety, so I willingly joined the team. At first glance the work seemed to be more like a translation than a study. However taking more careful look at it, I realized I could learn more. I could better understood the current status of the commercialization of GMOs in Korea in particular and noticed how much knowledge and awareness that Koreans will need in order to accept them with an easy mind. The work I have done gave me insight not only to better language ability but also to better ways to promote successful release of GMOs. To hear the research resound in my ears and to taste them in my mouth would be an almost unimaginable pleasure.

I have translated the words of this book, but my hope is that in the following pages I have also translated something of my ardor and awe in reading them.

Kim, Min Sik

***

Translating the book was much difficult than I first thought when I excitedly decided to take part. Since the book was very scientific and about the subject I scarcely knew anything, the GMOs, I struggled with every word. Of course it took much longer time than expected, and I was losing interest. I would like to thank Nawoo who managed the whole translation project and other members as well for helping me to keep up till the end and get to know this feeling of achievement after finishing the whole translation. Even though the content was very difficult, I could learn many interesting facts about the GMOs, and I think because of its difficulty, completing this project means more to me. I'm very grateful to have this opportunity of translating the book about field I am interested in, biology, and also thankful to all the members who worked together for publishing the translated version of book.

Lee, Chae Eun

***

I can still vividly remember the day when I first encountered the field of bioengineering. I was watching TV when I chanced upon a cloned dog named, Snuppy. Until then, I had not once heard the word "bioengineering". Yet, the fact that Snuppy was identical to its mother in every way completely shocked and fascinated the then ten-year- old me.

Since then, my interest in bioengineering has become much broader. Nevertheless, as I continued studying more about bioengineering, I came face-to-face with many dilemmas that existed in the field.

There were several debates going on about ethical issues and safety problems. At times, I did not know what to believe or what ideas to hold on to. I even thought about stopping my pursuit of bioengineering.

Before I could come up with a solid decision on what standards to follow and what practices to dispense, I had to read many books, articles and the writings of field experts. Overtime, I arrived at the certainty of wanting to pursue bioengineering and decided my own standards; but, I also know that I could not come to this decision without the assistance of books that provided me with diverse information.

Biotechnology on a Table was one of the many excellent books that provided plethora of information about the biologically engineered food that we eat and was one of the books that affected my conviction. Through these books, I have gained insight that our need for biologically engineered food is increasing more than ever. Moreover, we cannot simply ignore it because of its safety problems and ethical issues. Rather, I believe that we should study more about biologically engineered food and directly face its advantages and disadvantages to make proper decisions on biologically engineered food issues.

This book helped me to rethink about biologically engineered food and the standards that I had for them. In addition, I acquired much interest and information about biologically engineered food. I was impressed and I wanted to share this with others who had not yet found their own conviction on bioengineering. However, the fact that the book was written in Korean made the range of possible readers very narrow.

While I was translating this book, my dream was to share this excellent book with global readers. Also, I truly hope that this book will help many readers, the way it has for me, in making their own standards on bioengineering.

Park, Jieun

***

Biotechnology can be a key to the solution of many problems that mankind is currently facing. For example, scientists partially solved global food problems by producing genetically engineered crops. Despite numerous advantages that biotechnology can provide, people are controversial over it. Actually, I am still unsure whether biotechnology is a proper means to solve such problems. I wanted to find out whether biotechnology is suitable for solving such social issues. Consequently, I became interested in biotechnology and how it can solve human problems. I read some books regarding biotechnology, but I had little information about it. To gain more information about biotechnology, I decided to translate this book and by doing so I could learn a lot about it; I could attain balanced and objective information concerning biotechnology. The most difficult thing in translation was to translate many technical terms into proper English equivalents. I had to study all kinds of related concepts to understand and translate such vocabularies. However, I could learn more about various areas that biotechnology covers other than GMO crops such as blue roses, decaffeinated coffee, or mine-detecting plants. People who blindly oppose to biotechnology should learn about both benefits and harmful effects that biotechnology might bring and how safe or dangerous such technologies are. Translating this book was a lucky opportunity for me, who always desired to learn about biotechnology. I want to recommend this book to those who extremely support or oppose biotechnology. This book will help to develop discerning perspectives on that area.

Suh, Juhyung

***

Many experts claim that within a century humans would suffer from overpopulation, lack of resources and scarcity of food. The new conclusion from biologists was the genetically modified organisms, GMOs. Since then, there were many social disputes whether those unnatural organisms were safe. This book tries to show how those biotech crops are made, assured and commercialized with explanations and experiments; this book is a guideline to understand the genetically modified organisms.

8 months ago, when I started to translate this book, I could never imagine how successful and meaningful this work would be. It was hard to accurately change the biological words in to English and especially minimizing the gap between two languages was more difficult. However, the fact that I ever had a chance to let English speakers read this book was grateful.

Woo, Sunyoung